NATURE'S WAY

A BOOK OF ESSAYS ABOUT NATURE

A. L. (Tony) Gennaro

Professor of Biology
Director of the Natural History Museum
Eastern New Mexico University-Portales

*To Alice +
Norman.
Hope you
enjoy New
Mexico*

23 Aug 07

Cataloging-in-Publication Data

Gennaro, A.L., 1934-
    Nature's way: a book of essays
about nature/by A.L. (Tony) Gennaro.

    Includes bibliographies.
    Illustrations by author.

    1. Animal behavior. I. Title

QL751 G46 1996 591.5dc 96-75559
    ISBN 0-9648431-0-2 (pbk)

Marnaric Press, Inc.
P.O. Box 966
Portales, NM 88130-0966

Printed in the United States of America

Dedicated to my wife,
Marjorie

PREFACE

Humans strive to dominate Earth, and at the same time, they demonstrate little regard for life that is not human. Along with bacteria, viruses, and insects, humans are reaching a status of dominance. But, unlike those forms of life, humans can reason, and the ability to reason should make them aware that their survival depends on other species. Philosophers and naturalists encourage humans to step outside their culture into nature and apply ethics there in the same way ethics are applied among humans (the golden rule). However, ethics toward nature exists only when knowledge and appreciation prevail. It is my intent, through concise, easy-reading essays on interesting subjects, to provide that knowledge and appreciation.

A. L. (Tony) Gennaro
Portales, New Mexico

ACKNOWLEDGEMENTS

I have many individuals to thank for their assistance in the preparation of Nature's Way. One is Gary Pfaffenberger, my colleague at Eastern New Mexico University (ENMU). Gary and I worked together to maintain accuracy of information. Roger Garrett, KENW-FM Production Director, assisted me in converting complex subjects from the literature into simple, concise readings. Other individuals who were extremely helpful include students, or former students, of ENMU. These students include Leslie Reese, Bruce Corley, Lisa Marks, Brooks Lane, Jana Houston, Shay Boatman, Allison Estrelle, Dana Faris, Tammy Boden, Martin Flores, Terri Morris, Susan Bengiovanni, Eric Frey, and Michele Hallisey. Other students, such as Tonya Montessi, Kathie Haman, and Kathy Morgan maintained a system of order among reference materials and written documents. Deborah Pannabecker organized the manuscript for publication. May Dowlin, Pat Fleming, and Frances Fleming proofread the manuscript. I am truly indebted to Karen Pacheco and Ilsa Jorgensen for their constructive criticisms. Finally my wife, Marjorie, supported me during the preparation of Nature's Way, and she assisted me with the final edit. I thank her, my true companion, for making the publication of Nature's Way possible.

ABOUT NATURE'S WAY

Nature's Way essays are published in several New Mexico newspapers, and they are part of the author's lecture series in the course entitled, "Wildlife Biology," at Eastern New Mexico University (ENMU), Portales. All 80 episodes of Nature's Way were recorded in the KENW-FM studio as scripts under the direction of Duane Ryan, Director of Broadcasting, for national distribution. The author's co-host on the radio shows is Gary Pfaffenberger, who is a Professor of Biology in the Department of Biology at ENMU. The author and Gary Pfaffenberger have published several articles on biology in scientific journals.

ABOUT
THE AUTHOR

A. L. (Tony) Gennaro is a native of Raton, New Mexico. He received a Bachelor of Science in zoology and a commission in the United States Army at New Mexico State University, Las Cruces. Tony earned a Master of Science and Doctor of Philosophy at the University of New Mexico, Albuquerque. He is a naturalist with a particular interest in terrestrial ecology.

Tony Gennaro is active in scholarly research and teaches biology. Preceding active duty in the military where he received an Honorable Discharge as Captain, he taught at Las Cruces High School, Las Cruces, New Mexico. Tony's teaching continued at St. John's University, Collegeville, Minnesota and continues at Eastern New Mexico University, Portales, where he is a Professor of Biology and Director of the Natural History Museum. Background for essays in Nature's Way comes from Tony's research, classroom teaching, experiences with students, and from the knowledge he gained as a writer and host of two television programs and a radio show.

CONTENTS

WOLF VOCALIZATIONS

Howls of wolves are very distinctive. They are prolonged, deep, wailing, and are not uttered by any other kind of animal. Those who have heard them in the wild say they are the most dreary sounds perceived by humans. We can hear them up to 10 miles from their source of origin, but wolves can hear howls from much greater distances. Experts estimate that howls cover an area of approximately 300 square miles.

The frequency of howls varies seasonally. They increase in number from the end of summer and attain their highest levels in midwinter during the height of the breeding season. After that, they diminish in amount, but continue to occur.

Wolves howl to maintain contact with other members of the pack. The pack consists of a breeding pair—the alpha male and his mate, the alpha female—as well as their nonbreeding offspring from several successive generations. Maintaining contact with each other is of vital importance because wolves function as a family unit during hunting, care of the pups, and in other ways. Therefore, togetherness must be maintained, and the howls are their means of cohesion.

Howls are especially heard following a hunt when members of the pack separate from one another. The rendezvous is usually initiated by howls of the alpha male. Soon, other members of the family gather around him. Howls are uttered by wolves returning to the rendezvous site, as well as from adults and pups at the site.

Once gathered in their territory, an area the pack claims as their own, their howls will then take on other functions. For one, they warn neighboring wolves to respect territorial boundaries. This allows the pack to carry on its activities without disturbances from nonkin.

Wolves are definitely social animals, with an attraction for kin, and they function primarily as a social group. Of special interest to us is, how are the social needs of wolves met by the domestic dog which is a direct genetic descendent of the wolf?

Since domestic dogs have limited or no access to other dogs, they relate socially to humans in the same way their ancestors related to wolves. Dogs continue to howl as they did in the past, sometimes in response to technological devices, such as high tones of musical instruments, sirens, and whistles, as well as the howls of other dogs. We wonder why. Are these dogs attempting a rendezvous? Are they requesting recognition of territorial boundaries? And, how about the dog who initiates the howling, seemingly without a response to other howls or howl-like sounds? Is this dog demanding a return howl from kin to rendezvous, to defend territory, or is this dog howling to seek companionship, closeness, sociological union? Is the dog searching for social unity with an ancestral wolf, with other dogs in the neighborhood, or with its human keeper? Unfortunately, these questions may never be answered.

BIBLIOGRAPHY

Allen, D. L. 1979. Wolves of Minong: their vital role in a wild community. Boston: Houghton Mifflin.

Hall, R., and H. S. Sharp. 1978. Wolf and man: evolution in parallel. New York: Academic Press.

Turbak, G. 1987. Twilight hunters: wolves, coyotes, and foxes. Flagstaff, AZ: Northland Press.

THE SAGUARO

Certain images come to mind when the word saguaro is mentioned. These images include a giant cactus of Arizona; a plant which resembles a tall, slim man standing on the desert floor with his arms bent skyward; a tall cactus with the top resembling a pitchfork; or the typical cactus in a desert movie. All these comments describe the tall giant in the Arizona desert.

This giant is the state flower of Arizona, a good choice by the Arizonians. This cactus inhabits southern California and northern Mexico, but it is most abundant in Arizona, where a national monument has been established 17 miles east of Tucson solely for the protection of the saguaro.

Here are some facts about the growth and height of the saguaro. Some reach 58 feet. That is about six stories high. However, most reach 40 to 50 feet, and that is when saguaros are about 150 to 175 years of age. As might be expected, this immense size does not occur overnight. The height of this cactus, with respect to age, is about one-fourth inch the first year, three feet by the twenty-first year, and seven feet by the time saguaros are 30 years old. Now, that is slow growth in anyone's estimate. Saguaros reach reproductive maturity at about eight years and then develop their first arms when they are 15-feet tall and 75 years of age.

Saguaros are also heavy, reaching up to 2,500 pounds or more. That kind of weight can only be supported by several wooden rods which are evenly spaced on the inside periphery of the plant. Some of these rods are two inches in diameter, and they run the entire length of the trunk and arms. The weight of the saguaro results from the weight of these rods, plus water which the plant stores.

Water is a very important component of this plant, and one wonders how such a huge structure acquires this liquid. The root system does the trick. It consists of many shallow roots radiating out from the base of the trunk. These roots are greater in length than the height of the plant, and they quickly absorb any water that penetrates the soil following sparse rainfalls of the saguaro habitat. The saguaro is well adapted to store water, hence the accordion pleats on its surface which expand in times of plentiful rainfall. Although this water is unpalatable to humans, rodents and rabbits consume it when they chew saguaro tissues.

There are many other ways in which saguaros are important to desert dwellers other than a source of water. For example, flowers of the saguaro open at night and provide nectar to insects and the long-nosed bat which pollinate the saguaro flowers. Birds nest in cavities which they excavate in the saguaro. Fortunately, the cactus secretes a substance that lines the nest cavity. This lining acts as a sealer which prevents plant juices from leaking into the cavity. Also, this lining prevents insects from penetrating plant tissues beyond the lining of the nest cavity. Therefore, nesters cause no harm to the saguaro.

Certainly most of us have no intent to harm the saguaro. We are concerned about the welfare of these magnificent treasures of the desert, and we are grateful to those individuals who had the foresight to establish means to ensure their survival.

BIBLIOGRAPHY

Carlson, R. 1954. The flowering cactus: an informative guide. New York: McGraw-Hill Book Company, Inc.

Leese, Sir O. 1973. Cacti. London: Tribune Books.

Wild, P. 1986. The saguaro forest. Flagstaff, AZ: Northland Press.

PAPER WASPS

Nests of paper wasps are easily recognized. They frequent porches, eaves, or patios and resemble champagne glasses hanging upside down by their stems from the underside of these structures. The cavity of the glass-shaped nest is packed solidly with six-sided wasp-rearing chambers. They are all positioned parallel to each other, and their open ends face downward and are about even with the rim of the nest cavity.

A wasp nest usually sparks a familiar reaction from people. One comment includes, "Quick, get the broom, and I'll knock that nest down." Such responses are necessary if individuals who are allergic to wasp stings are nearby. Otherwise, removal may be uncalled for. Many people have observed wasps on their porches summer after summer and have never had to dodge an attacking wasp. In fact, wasps are about as likely to sting humans as they are to sting a fence post.

Nests are easily destroyed because they are composed of paper which wasps construct from chewed plant fibers mixed with fluids from their mouth. These fibers are obtained from materials such as vegetation, unpainted boards, and cardboard boxes. The nest building process is initiated by two to ten females, usually sisters, which received sperm from males the preceding fall season. One of these females will become queen of the nest and will be the only one that lays eggs in the rearing chambers. Other females, which also contain sperm and are capable of laying eggs, serve as workers, in that they hunt for insects and nectar and feed the queen, as well as the larval wasps which hatch from the queen's eggs. If the queen dies, one of the other females will replace her and continue the egg-laying process.

Female wasps are an organized group, displaying rank and order. The queen is the top-ranking individual. She, the alpha, dominates a beta female, who in turn dominates another female, and that female dominates another in that order on down through the line of females on the nest.

Female wasps display individual waspalities (that is wasp talk for personalities). Some individuals perform special chores. Others are rather laid-back and do little or nothing. Some are aggressive; others are not.

Adults emerge from nest chambers in about 32 days. As this occurs, another egg is laid within that cell. The queen fertilizes early hatches with sperm she received the previous fall. These will hatch females. At the end of the summer, by an inherent capability, the queen prevents sperm within her body from fertilizing eggs she lays. These will hatch males.

With all these wasps of different ages and sexes, the nest is a site of confusion near the end of summer. Some young females assist the adult females with chores. Others just sit and groom themselves. Males remain idle most of the time and get in the way. But, at the end of the summer, as all wasps gradually leave the nest, males finally get their assignment, and that is to mate with as many females as possible before these females enter hibernation. This mating process loads the females with genetic material and prepares them to establish a new generation of wasps the following summer. Having accomplished their mission of sperm transfer, the role of the males is over. They die soon after mating.

BIBLIOGRAPHY

Evans, H. E., and M. J. West Eberhard. 1970. The wasps. Ann Arbor: University of Michigan Press.

RATTLE- SNAKE VENOM

Rattlesnake venom is well known. Humans think of it as a liquid of death which is injected into tissues during a snakebite. However, to rattlesnakes venom is an instrument of life. The venom's primary function is not to kill attackers but to immobilize prey for ingestion.

When a rattlesnake punctures flesh with its fangs, venom is squeezed out of glands situated in the head of the snake. The venom is forced through two hollow fangs into tissues of the victim. The snake then waits until the struggles of the prey terminate before swallowing it.

Would venom have been useful to early native American buffalo hunters? A dip of an arrow or spear into a venom-like substance, not toxic to the hunter of course, would have caused the buffalo to fall to the ground quickly and remain there harmless, immobile, and on the edge of death. With arrows alone, the hunter had to wait for the buffalo to drop after it lost a sufficient amount of blood or deal with an angry, wounded buffalo. No wonder nature provided venom for rattlesnakes to ease the difficulty of immobilizing prey for consumption.

Generally, snake venom is of two kinds. One is neurotoxic and affects respiratory or heart action. The other is hemotoxic and destroys blood vessels and blood cells. Venomous species often have both kinds of venom. In some groups, such as cobras and coral snakes, venom is more neurotoxic than hemotoxic. In rattlesnakes, on the other hand, venom is more hemotoxic than neurotoxic.

Venom may remain effective for many years even when isolated from the body of the snake. For example, venom will remain active for 50 years in a dried condition and 20 years as a liquid. Therefore, it is not surprising that an individual who is scratched or punctured by the fang of a dead rattlesnake still experiences the effects of its venom.

Venom can cause serious damage to the victim, especially to tissues underlying protective layers of the skin. This should discourage individuals from treating bites of rattlesnakes with the out-dated cut and suction method which involves oral suction of venom from the bite wound. Individuals with cracked lips, bleeding gums, intestinal ulcers, or other open lesions could expect trouble if they sucked venom from a snakebite wound.

Then, there are those who wish to possess a pet rattlesnake (that is, a harmless one). The first thought is to remove the snake's fangs. However, removal of the fangs will not render the snake harmless. Damaged fangs are quickly replaced, a process which is necessary because fangs of a rattlesnake easily break off during initial contact with a victim. This replacement process and many other adaptations of rattlesnakes ensure the flow of that vital fluid of life—venom. Without it, rattlesnakes would have to wait for their prey to bleed to death after stabbing them with venomless fangs.

BIBLIOGRAPHY

Boys, F., M.D., and H. M. Smith, Ph.D. 1959. Poisonous amphibians and reptiles: recognition and bite treatment. Springfield, IL: Charles C. Thomas.

Burton, R., M.A. 1978. Venomous animals. New York: Crescent Books.

Carr, A. 1963. The reptiles. New York: Time Incorporated and the Editors of Life.

Davidson, T. M., M.D., and S. F. Schafer. April 1989. Rattlesnakes: the animal and the venom. The Physician and Sports Medicine 17(Part 1 of 2).

Gans, C., and R. B. Huey. 1988. Biology of the reptilia: defense and life history. New York: Alan R. Liss, Inc.

Shaw, C. E., and S. Campbell. 1974. Snakes of the American west. New York: Alfred A. Knopf, Inc.

DOMESTICATION

All domesticated animals are descendants of native species taken by humans from the wild and bred. From the progeny of each generation, only desirable offspring were selected for further breeding. Those selected displayed traits which fulfilled human needs. By choosing specific offspring for breeding, certain genes were retained in the genetic line; whereas, undesirable genes were denied access to succeeding generations. The result has been breeds designed to provide labor, companionship, food, clothing, and other products for human cultures.

Dogs are an example of domestication. They originated 12,000 years ago in Iraq. They were derived from wolves to serve humans in work, play, and sport. The appearance of certain breeds of dogs makes it quite difficult to believe that the wolf is their ancestor. For example, a Chihuahua certainly does not resemble a wolf. The reason is many of the wolf genes were eliminated. On the contrary, it takes less imagination to envision the German shepherd as a descendant of the wolf because fewer wolf genes were eliminated.

Individuals with ethical attitudes toward preserving wildlife wonder just how far the German shepherd is removed genetically from the ancestral wolf. Furthermore, some of them might consider it possible to reestablish native wolves again through selective breeding of the German shepherd. Skeptical scientists consider that process impossible because native wolf genes, once lost in the domestic genetic line, cannot be restored. This is true for all domestic strains. However, some individuals remain optimistic.

The Tarpan, from the grasslands of southeastern Europe, is thought to be the ancestral wild horse of all domestic horses. It was moderate in size, gray colored, with a black mane and tail. It became extinct in Russia during the 19th century. Domesticated horses with conspicuous Tarpan-like features were bred selectively to develop a horse which would resemble the original wild European species. Horses from those matings now reside in various European and United States zoos, and they do resemble the original native Tarpan.

However, most authorities agree that once domesticated, always domesticated. In other words, if a domesticated breed should return to the wild, it cannot be a wild species again. The word, feral, has been coined for such free, domesticated animals. Mustangs are good examples. They are frequently called wild horses, but they are really feral horses.

Besides mustangs, other domesticated animals are familiar to us in their feral way of life. The two most noteworthy are pigs and cats. Both survive nicely without human assistance.

Domestic pigs originated from the native wild boar of Europe and Asia. It was probably domesticated as early as 8,000 years ago in Thailand. About 5,000 years later, Polynesians introduced the domestic pig into Hawaii. During the 16th century, the domestic pig was introduced into southeastern United States by the Spaniards. Later, European wild boars were introduced into the United States for sport. Now, the free-running pig in the United States is either feral, an introduced wild exotic from Europe, or a combination of both.

Cats originated about 4,000 years ago in Egypt from the wild cat, *Felis silvestris*, of Africa and southeast Asia. *Felis silvestris* (from which the name Silvester the cat originated) is still a wild species in Europe, Africa, and Asia and is slightly larger than the typical domestic breed. This wild species has long, thick, grayish fur with a few thin black bars running from the forehead, between the ears, to the base of the neck. Another single black stripe continues to the base of the tail. The tail has several black rings around it and is black tipped.

I personally observed most of those characteristics in an unfriendly, unapproachable feral cat which frequented my neighborhood for about four years. I called him Tom, and I thought if any feral cat could make it in the wild Tom could, but I admit I was very concerned about his comfort on cold winter evenings.

Tom's fate raises a point of concern for all domesticated breeds. Some, like feral Tom, have a chance to survive in the wild. Others, however, are highly dependent on human care. Does this mean that domesticated animals, even feral ones, are the liability of humans who took them from the wild in the first place? Should humans insist that domesticated breeds be used effectively and responsibly for the reasons they were bred? Seemingly, these are questions that we should consider.

BIBLIOGRAPHY

Briggs, H. M. 1949. Modern breeds of livestock. 4th ed. New York: Macmillan Publishing Company, Inc.

Nowak, R. M. 1991. Walker's mammals of the world. 5th ed. Vol. 2. Baltimore: The Johns Hopkins University Press.

SEA SNAKES

Snakes in the ocean may seem out of place; however, sea snakes do live there successfully—all 50 species of them. These snakes typically occupy shallow oceanic waters off the coasts of several continents. Sea snakes have a size typical of some species of land-dwelling snakes, that is, three to five feet in length, with some species reaching a length of nine feet.

Another feature is held in common between sea snakes and land-dwelling snakes, both kinds breathe with lungs. Sea snakes do not seem to be hindered by this air-breathing habit. In fact, sea snakes can hold their breath long enough to dive 20 to 30 feet to capture food, a distance from the top to the bottom of a three-story building.

However, unlike land-dwellers which breathe with nostril openings on the front of the head, sea snakes breathe with nostril openings situated on top of their head. Like dolphins, porpoises, and whales, this air inlet on the top side of their head allows convenient air intake when snakes come to the surface. Also, nostrils of these snakes have valves which close when the snakes submerge to swim.

Sea snakes are effective swimmers. These animals move their bodies from side-to-side to propel themselves in water in the manner that most land snakes move their bodies from side-to-side to propel themselves on land. The motion of sea snakes in water is aided by a tail which is flattened from side-to-side. A flat tail, in comparison to a round one, enables more surface area of the snake's body to push against the water to enhance the forward motion of the snake. And, unlike most animals adapted to swimming in the ocean, sea snakes can swim backwards.

Because of their aquatic way of life, species of sea snakes that reproduce in the water must give birth to live young. These young must adapt quickly to the ways of their parents to avoid drowning. Some species lay eggs, but in all cases these are laid on land because the developing embryo in an egg would quickly drown if the egg were laid in water.

Equal in magnificence to breathing, swimming, and bearing young is the ability of sea snakes to capture prey. Prey of sea snakes, which includes fishes and eels, do not lend themselves to being squeezed to death as do prey of nonvenomous land snakes. These land snakes kill prey by constriction in that they wrap their body around the victim and squeeze tightly enough to prevent it from inhaling air into the lungs. However, prey of sea snakes breathe by gills. Therefore, squeezing would not be an effective method of killing. Furthermore, prey of sea snakes are highly mobile. If these prey are not killed quickly, they would easily escape into the depths of the ocean. Therefore, the only logical means for prey capture is the use of a very toxic venom. That is why the venom of sea snakes is more toxic than any venom known, except for one species of land-dwelling snake in Australia.

The procedure for venom injection is similar to that used by land snakes. The venom of the sea snake is injected into the muscle of prey through fangs situated in the front of the snake's mouth. After injection, the sea snake backs off and waits for the prey to become immobilized before swallowing it.

A predator with venom as toxic as that of the sea snake would be expected to be quite ferocious; however, this is not the case. Sea snakes are the most harmless of poisonous snakes. These snakes rarely molest bathers, and fishermen must be careful when removing snakes from fish nets. Apparently, sea snakes only strike at humans when humans accidentally kick or step on them in shallow water.

BIBLIOGRAPHY

Mertens, R. 1960. The world of amphibians and reptiles. New York: McGraw-Hill Book Company, Inc.

Pope, C. H. 1955. The reptile world: a natural history of the snakes, lizards, turtles and crocodilians. New York: Alfred A. Knopf, Inc.

GRIZZLY BEARS

Kodiak bears, grizzly bears, and brown bears are all the same species—*Ursus arctos*. Kodiak bears are found in coastal Alaska; grizzly bears inhabit inland Alaska, western Canada, and western contiguous United States. Brown bears reside in Europe and Asia.

From among those three groups of bears, it is the grizzly bear which is of concern here.

The name, grizzled, was given to these huge mammals because the long hairs on their back and shoulders are often frosted with white, thus giving these bears a grizzly appearance. Grizzlies cannot be confused with any other bear in the U.S., not even the black bear which shares its range.

Black bears with the scientific name, *Ursus americanus*, are very distinct in appearance and habit from grizzlies. For example, as compared to black bears, grizzlies have a prominent hump on the shoulder area, snout that rises more abruptly into the forehead, longer pelage and claws, longer body, and a weight 350 pounds heavier. Unlike black bears, grizzlies cannot climb trees, and grizzlies occasionally eat black bears.

Food of grizzlies is truly omnivorous, meaning a very broad diet. In addition to black bears, grizzlies consume edible vegetation, berries, roots, rodents, and salmon. In the Canadian Rockies, grizzlies eat moose, elk, mountain sheep, and mountain goats. In northern parts of their range, where grizzly bears hibernate, enough food must be consumed during active times to sustain them during their winter sleep.

This hibernation or winter sleep, however, is only partial. Grizzlies show a marked reduction in heart rate and a slight drop in temperature, but they do not enter a deep sleep like other true hibernators. Grizzlies are easily aroused from winter sleep and even give birth during hibernation.

Development of embryonic grizzlies begins as soon as fertilized eggs implant in the mother's uterine tissues. Although grizzlies mate from May to July, egg implantation is delayed until October or November. Young are born from January to March, and they are naked and blind. Each newborn weighs only one or two pounds and can fit into the human hand. Young in this condition are definitely in need of constant maternal care. In fact, the closeness of mother and cubs is so intense that cubs may remain with their mother until age four years.

It would seem that all is great in the life of a grizzly bear, except for one thing. Grizzlies and humans prefer the same home and recreational sites. In addition, both species prefer the same kinds of foods, especially livestock. Because of this preference for livestock, grizzly bears have been exterminated in many states.

As a result of grizzly bear-human conflicts, numbers of bears have been drastically reduced. For example, in the contiguous U.S. there were 100,000 grizzly bears during the early 19th century. At the present time there are fewer than 1,000 grizzly bears in that area. Western Canada and Alaska have about 50,000 grizzlies. Grizzly bear numbers are decreasing, and this reduction is the concern of conservation groups. All populations of grizzlies in the contiguous U.S. are now listed as threatened, and numbers of grizzlies will likely reach a threatened status in Alaska because of explorations in that area.

BIBLIOGRAPHY

Nowak, R. M. 1991. Walker's mammals of the world. 5th ed. Vol. 2. Baltimore: The Johns Hopkins University Press.

Tony Gennaro 94

COMMENTS ON
SHARK ATTACKS

All of us are interested in shark attacks on humans. The reason may be the method in which sharks attack. Each of the shark's many teeth is like a double edged, sharp, pointed knife blade. These teeth, along with the muscular power of shark jaws, interact violently with soft human flesh. The result is deep gashes, removal of chunks of human flesh, or dismemberment of body parts and bloody sea water. Certainly this scene is more dramatic than when humans are bitten by deadly scorpions, spiders, and snakes. Perhaps the reason is that those critters cause no immediate trauma to human tissue, and they do not consume human prey following attack.

Whatever the reason, the words "shark attack" perk up ears. As a result, sharks have attained the title of human-eaters of the seas. Are sharks deserving of that title? The answer depends on whether humans are an important item on the shark menu, or if humans are attacked for other reasons. For example, what is the influence of water temperature, bad weather, presence of murky waters, presence of shallow waters, rough seas, or calm seas on the rate of shark attacks on humans?

Answers to these questions can only be obtained from scientific research involving controls, experiments, and the application of statistics. However, such experiments are virtually impossible because humans are unwilling to volunteer as potential victims for such tests. How would sharks endure such experiments? The fact is that interesting as they may seem, experiments designed to answer questions about the nature of shark attacks are very difficult to establish.

An easier but less scientific way to answer such questions is to examine written records of shark attacks. Relationships between shark attacks and environmental factors may then be researched. However, these relationships, typically called correlations, should be treated with caution. Written records may indicate a strong correlation between certain water conditions and shark attacks, but these correlations do not indicate cause and effect. As another example, increase in shark attacks may increase at the same rate as a certain factor increases, but that does not mean that the factor in question is the cause of the increase in attacks. Cause and effect must be proved. Despite these obstacles, one courageous individual, H. David Baldridge, studied shark attacks from written records.

Baldridge was an officer scientist in the United States Navy at the time of his investigation. He analyzed over 1,600 recorded instances of shark attacks on humans since the first recorded incident in 1580. Data from those instances were compiled by the U. S. Navy in cooperation with the Smithsonian Institute.

Here are some of Baldridge's findings. From among 339 or more species of sharks, only 45 species are attackers. In addition, few humans have been eaten alive by sharks in the strictest sense of the term; therefore, it may be a gross overstatement to refer to every attacking shark as a human-eater. Baldridge stated further that 50 to 75% of shark attacks on humans had nothing to do with the hunger drive of the shark. Baldridge based this interpretation on the reported presence of gashes, slashes, or cuts on the victims inflicted by a sideways movement of the teeth on the open upper jaw of the shark. Such injuries are not caused from a feeding drive. A feeding drive is expected to cause teeth marks in flesh made from both the upper and lower jaws of the shark, with flesh missing from the bite area.

Despite his laborious efforts, Baldridge could not detect correlations between abiotic (nonliving) factors and the rate of shark attacks on humans. He did, however, discover one very important biotic (living) factor which was correlated with the rate of shark attacks—bathing pressure. Baldridge stated that the world rate of shark attacks on humans is more likely a function of human bathing pressure (that is, the number of people in the water). This means that as the number of people in oceanic waters increases, the rate of shark attacks increases, but again, there is no proof that these kills are a result of sharks seeking humans as food. In fact, humans may only resemble shark prey, such as seals, when humans are in the process of spear fishing and surfing. Perhaps humans are invading space occupied by sharks and are unknowingly provoking them to attack.

Finally, Baldridge made a strong point when he stated that sharks have been on Earth long enough to watch the dinosaurs come and go. Why should humans, in their very brief moment on this planet, attract the attention of sharks only momentarily? If humans were truly a food choice of sharks, the number of human deaths from attacks would certainly be higher than 15 to 20 per year worldwide. As Baldridge pointed out, there are 17 human deaths from bee stings in the U.S. each year. Why should sharks be so alarming? In another comparison, about 40,000 humans are killed in the U.S. in automobile accidents annually, yet drivers demonstrate little fear of operating an automobile. Perhaps we have overrated sharks as a serious threat to our lives.

BIBLIOGRAPHY

Baldridge, H. D. 1975. Shark attack. New York: Berkley Publishing Corporation.

Exotics of

the Americas

Certain species of plants and animals are not native to the New World (the Americas), meaning they did not originate there. Since these species originated elsewhere, such as Europe, Asia, and Africa, they are said to be exotic to the New World.

The tumbleweed is an exotic. This plant was imported with desirable seeds via ships from Russia, hence its other common name, Russian thistle. The first evidence of this plant was on a farm in Homme County, South Dakota in 1877. Tumbleweeds thrived in prairie sod broken from cultivation, as well as in overgrazed areas. By the turn of the 19th century, these hardy plants tumbled from the Dakotas to the Pacific Coast, scattering seeds along the way. Now, Russian thistle grows in a major part of the central and western U.S. with scattered populations east of the Mississippi River.

Tumbleweeds arrived in the U.S. later than three unwanted rodents which accompanied humans during their travels by ship to the New World. These rodents include house mice, Norway rats, and black rats. The house mouse originally came from an area extending from the Mediterranean region to China. The Norway rat came from China and the black rat from the Malaysian region. At the present time, the house mouse and Norway rat occupy all areas of the U.S. where humans dwell. The black rat, on the other hand, inhabits southern parts of the U.S. near coastal areas. All three species destroy valuable human products and transmit diseases.

One such disease is bubonic plague which is caused by the plague organism, *Yersinia pestis*. This organism is transmitted from fleas to mammals when fleas obtain their blood meal from mammals. The plague organism is responsible for the "Black Death" in Europe which killed 25 million humans during the 14th century. Plague was recognized for the first time in the New World in San Francisco in 1900. It is thought that the plague entered that city when infected rats escaped from a ship from the Orient. Soon after the infected rats arrived, plague organisms spread to native rodents and rabbits of the American Southwest. Since 1900, there have been over 1,000 cases of human plague resulting in more than 700 deaths in the U.S.

Fortunately, exotic birds have not been as detrimental to human populations as exotic rats. There are three common avian exotics in the U.S. Two are the European starling and house sparrow (or English sparrow), both of which were introduced into the eastern U.S. during the 19th century. At the present time, both species occupy all parts of the U.S. The third avian exotic is the pigeon which was domesticated from the wild rock dove by Egyptians as early as 3,100 B.C. Many centuries later, the domestic pigeon was introduced into the U.S.

All of the aforementioned exotics were either introduced accidentally (tumbleweeds, house mouse, Norway rat, black rat, and the plague organism), or they were introduced intentionally (house sparrow, European starling, and domestic pigeon). Many species of insects were accidentally introduced to the Americas, but a large number of game species were introduced intentionally for hunting purposes. The question is, who is responsible for these accidental or intentional introductions of exotics? Humans. That is why these exotics are closely associated with humans and have been seen by humans at one time or another.

What about humans? Are they exotics of the New World? The answer is yes. Most anthropologists agree that humans crossed from Asia (Old World) to Alaska (New World) by a land bridge across the Bering Strait. Those humans (later called American Indians) eventually dispersed over the Americas. The first widely accepted evidence for the earliest humans in North America was established when artifacts of humans 11,500 years in age were discovered near Clovis, NM. Since the Clovis find, human artifacts of similar age have been discovered at other sites in the Americas. More recently, discoveries of artifacts have led some anthropologists to believe that humans may have been present in the Americas much earlier than 11,500 years, perhaps between 20,000 and 30,000 years ago. However, most anthropologists welcome more independent finds of these earlier materials before agreement is reached on validity.

BIBLIOGRAPHY

Gill, F. B. 1995. Ornithology. 2nd ed. New York: W. H. Freeman and Company.

Nowak, R. M. 1991. Walker's mammals of the world. 5th ed. Vol. 2. Baltimore: The Johns Hopkins University Press.

Pearson, T. G. 1936. Birds of America. In three parts. Garden City, NY: Garden City Books.

Weber, N. S. 1977. Plague in New Mexico. Santa Fe: Vector Control Program. Environmental Improvement Agency, State of New Mexico Health and Social Services Department.

Tony Gennaro
45

NAVIGATION OF BIRDS

Facts about bird navigation are amazing. For example, the Mississippi kite, a falcon-like bird, departs from South America in the spring and migrates northward about 5,000 miles. These kites arrive in the United States about the same time each year, and there is good evidence that some of these birds frequently build a nest in the same fork of a tree which they used the previous year. That is precision, not only with timing, but with finding their way along a 5,000-mile route to a small target the size of a tree trunk. What methods do kites and other migratory birds use to navigate?

Some species of migratory birds use visual landmarks. These birds consistently follow routes along coastlines, rivers, and restricted land corridors, such as the Strait of Gibraltar and Central America. Use of landmarks is supported by the fact that none of these birds which follow corridors and coastlines ventures out over the ocean. To humans who navigate on continents, landmarks as guides are logical, but birds also use other ways to navigate which are not as easy to comprehend.

Tests to determine methods of bird navigation, other than the use of landmarks, make use of the behavior of birds confined to circular cages. When the time for migration arrives in the spring or fall, migratory species confined to circular cages orient or flutter in the direction which they would ordinarily migrate. For example, species which fly south in the spring, flutter toward the southern end of the circular cage as spring arrives. This fluttering behavior is referred to as migratory restlessness. The direction in which the birds flutter, the duration of migratory restlessness, and the times when this restlessness begins and ends have been the basis of many experiments.

One test in which these behaviors were used involved a study of the sun's position as a navigational tool. Migratory species held in circular cages with the sun in their view fluttered in a direction they would normally migrate. Mirrors were used to alter the positions of the sun, as viewed by birds. These alterations caused the birds to change their direction of flutter. By making these solar changes, researchers could predict which direction the birds would flutter, indicating that birds use solar navigation.

Similar experiments were conducted with star patterns to test the use of stellar navigation. When certain nocturnal species of birds were exposed to a star-lighted sky while confined to circular cages, they fluttered in the direction they would ordinarily migrate. This behavior supported the birds' use of stars as a source of directional information. Given the same pattern of stars in a planetarium as that which was present outdoors, the birds fluttered in the same direction as they did outdoors. As the pattern of stars in the planetarium changed, the birds fluttered in directions as predicted by the researchers.

The influence of the magnetic field was tested in a similar way. Migratory species in circular cages were exposed to changes in the magnetic field. Directions of flutter varied with changes in those fields, indicating magnetic field navigation.

All these experiments bring up one question. Which tool for navigation is inherited and which is learned? Answers were determined by studying hand-reared migratory species, again in the circular cages.

Discoveries from those studies indicated two things. First of all, birds inherit the means to determine at what time and how long it takes to travel during migration. In other words, if Mississippi kites were tested, it is expected that they would begin fluttering in a circular cage when their migration would be initiated, and they would flutter the length of time it would take them to travel 5,000 miles. Secondly, migratory species commonly inherit a sense of direction guided by forces of the earth's magnetic field. With experience, these birds learn the fine points of solar or stellar navigation. Additional navigational means are suspected, such as odors and wind directions among others, but they await tests to determine their significance.

Birds probably do not inherit recognition of specific features which characterize their home either on the winter grounds or the summer breeding grounds. Such an inherited program would not be effective because habitats are highly variable, not stable like magnetic fields. Therefore, inheritance may involve the selection of a general habitat, but birds must remember the specific features of a certain part of the nest tree if they use it the following year. It is likely that the bird's memory will distinguish a successful nest site from one that is unsuccessful. Obviously, these birds will return to the successful site which is a good choice on their part.

BIBLIOGRAPHY

Gill, F. B. 1995. Ornithology. 2nd ed. New York: W. H. Freeman and Company.

Welty, J. C. 1988. The life of birds. 4th ed. Philadelphia: W. B. Saunders Company.

Tony Gennaro
95

KILLER WHALES

Killer whales are truly spectacular to observe as they move their large, eight ton, black and white bodies through the water with the greatest of ease. Their movements are dolphin-like, and there is good reason for this similarity. Killer whales are the largest species of dolphins.

Killer whales are mammals. This means that these whales breathe air. Breathing is accomplished through a single nasal opening on the top of the whale's head. This opening is called a blowhole, and it serves as a passage for air to the lungs. Killer whales move their bodies up and down as they swim through the water. This movement allows them to take a breath of air through their blowhole as they expose that intake tube to air above the water. Fishes, on the other hand, breathe by gills; therefore, undulations from side to side suffice as their means of locomotion. Killer whales also have mammary glands just like land mammals. In most ways, these whales are as docile and gentle in appearance as any other species of dolphin, yet they are called killer whales. Most individuals wonder why.

Authorities state that the killer whale received its common name because it is the world's largest predator of warm-blooded animals; however, these authorities agree that killer whales are no more of a killer than any other predator. Robins kill worms, but they are not called killer robins. Swallows kill insects, but they are not known as killer swallows.

Killer whales are predators, however, and certain kinds of animals are on their menu. They include fishes, squid, dolphins, seals, octopuses, and sharks in that order of preference. Killer whales also enjoy the tongues of other whales which they obtain from whale calf kills. However, killer whales are best known for the tactics they display when they pursue seals.

Seals encounter killer whales in packs of 2 to 40 in number. Packs of whales cause such alarm among seals that seals have been reported to head for shore or even jump into boats to avoid capture. There is good reason for such behavior. Killer whales play with seals in the same way a cat plays with a mouse. Whales will toss a seal into the air several times before killing it. In one incident, a group of killer whales harassed a California sea lion for 45 minutes before ending its life. Seals are not even safe on beaches near the water's edge because killer whales will move out of the water a short distance and capture them. During exhibitions, trainers will frequently repeat that performance by luring captive killer whales out of their holding tanks headfirst onto the surrounding decking. Spectators on the front rows are splashed with sea water, but fortunately they do not find themselves in the mouth of the killer whale.

Seals also have little chance for survival on large ice floes, especially if they are on the edge of the ice. Killer whales dive deeply, rush to the surface, break the ice with their back, and flip seals into the water. Whales can break ice up to a meter in thickness.

Yet, despite this fierce behavior which killer whales display toward seals, these whales have never been known to harm a human. And, that is why spectators should never fear that a captive killer whale will harm trainers during performances.

Unfortunately, killer whales are in demand by humans for more than their ability to perform. People from various nations also use them as a source of meat and oil. The aforementioned prey choices of these whales were determined from the examinations of stomachs from 364 killer whales taken off the coast of Japan. Killer whales are also killed off the coasts of Norway and Greenland. During the 1979/1980 season, the whaling fleet of the Soviet Union took 916 killer whales. Fortunately, only 16 killer whales were reported taken in the 1982/83 season. Since that time, all commercial hunting of whales has ceased.

Perhaps the killer whale should be given a name less descriptive of its actions. How about the white spotted whale? After all, killer whales do have a white spot behind the ear.

BIBLIOGRAPHY

Burton, R., M.A. 1980. The life and death of whales. 2nd ed. New York: Universe Books.

Coffey, D. J. 1977. Dolphins, whales and porpoises: an encyclopedia of sea mammals. New York: Macmillan Publishing Company, Inc.

Gaskin, D. E. 1972. Whales, dolphins and seals: with special reference to the New Zealand region. Auckland, New Zealand: Heinemann Educational Books.

Martin, K. 1988. Giants of the sea. New York: Gallery Books.

Martin, R. M. 1977. Mammals of the oceans. New York: G. P. Putnam's Sons.

Nowak, R. M. 1991. Walker's mammals of the world. 5th ed. Vol. 2. Baltimore: The Johns Hopkins University Press.

INSULATING FUNCTION OF FUR AND FEATHERS

The Siberian tiger uses the color of its fur (hair) for concealment as it stalks its prey, and the red cardinal uses the color of its feathers to advertise its presence to other cardinals. These functions of mammalian pelage (fur) and avian plumage (feathers) are quite apparent. Yet, there is another function of these body coverings which is not so obvious.

Fur and feathers also conserve body heat which is generated by means of chemical reactions of metabolism. Conservation of this heat is necessary in order for mammals and birds to maintain a constant body temperature in cold environments. Heat conservation is accomplished because the structure of fur and feathers restricts the radiation of heat from the body. Because of this restriction, fur and feathers are good insulators of heat, but at the same time they are poor conductors of heat. Heat is not transmitted, or conducted, readily through fur and feathers.

Another good insulator is air. Air conducts heat less effectively than fur and feathers, and air is also used by mammals and birds to conserve heat. These animals trap air within their pelages and plumages in the following manner. When a mammal or bird is cold, a tiny muscle contracts at the base of each hair or feather. These muscle contractions pull on the bases of hairs or feathers and erect them, creating a fluffy appearance of the pelage or plumage that we commonly observe among animals on cold mornings. Erected hairs and feathers are the actual air traps. Trapped air forms a continuous blanket over the bodies of mammals and birds and prevents body heat from being transferred to the surrounding colder air of the environment. This amazing feat of nature has been put to use by humans.

For example, insulation in the walls of a dwelling traps air to form a blanket of insulation. This trapped air prevents heat within the building from transferring to the colder environment outside the building; therefore, the dwelling remains warm. Without this insulation, air moves freely in the walls and does not form a permanent, stable, insulating blanket.

Air blankets are also important to humans for their comfort during cold weather. However, because humans have relatively few hairs on their bodies, they must resort to various kinds of clothing produced from animal hides or woven materials which trap air. These include wind-resistant materials lined on the inside with down (soft feathers), wool of sheep (recall the sheepskin-lined bomber jackets), or various kinds of fibers which are effective air-trapping materials.

When mammals' hides are worn for warmth, they must be worn with the fur facing inward rather than outward because of the air trapping quality of hairs. There is enough fluff of the fur, especially soft fur, between body and hide to trap this air, providing the garment is not worn too tightly against the skin. Obviously, it is not functional to wear animal skins with the fur facing outward for warmth because muscles which erect hairs have long since disappeared with the life of the animal. Without those muscles, fur of bear, mink, muskrat, fox, or beaver will not fluff. The fur simply flaps in the breeze, without control. Without fluff, air is not trapped, and the furry garment simply acts as a windbreaker. Any garment worn with fur facing outward is likely done for fashion or to avoid discomfort because short hairs of some animal hides could irritate the tender skins of humans.

Humans do attempt to insulate themselves with an air blanket without use of animal skins and woven garments. Muscles attached to the few body hairs which humans possess contract as a response to cold. As each muscle contracts, it pulls on the base of a hair beneath the skin. This process erects an almost inconspicuous hair, and at the same time, it causes the skin surrounding the base of the hair to elevate. These areas appear as small bumps which humans refer to as "goose bumps." Although these bumps are nonfunctional, they are a warning that, without an air blanket, a warm shelter or protective garment is needed.

BIBLIOGRAPHY

Irving, L. 1968. Adaptations to cold, vertebrate adaptations: readings from Scientific American. San Francisco: W. H. Freeman and Company.

BLACK WIDOW SPIDERS

Black widow spiders are easily recognized. Females have a shiny black, pea-sized abdomen with reddish or yellowish markings on the underside. These markings are unique to species. A southwestern U.S. species displays a red hourglass. An eastern U.S. species has an hourglass, but the red triangles of the hourglass do not meet. A species in Florida has red spots surrounded by a yellow circle. Males lack this contrast of colors and are smaller than females. Males are brown and about one-half to one-third the female's size. Males are rarely seen except during courtship and mating.

At these times, males display a very subtle behavior. Insemination of females seems to be their main role, but he has a good chance of being consumed by the female (hence the name widow) during that process. In fact, female black widows consume 65 percent of males which mate with them. Females are obviously in control of the mating relationship.

The female is definitely the most conspicuous of the two sexes. She exposes her glossy black body without fear as she builds her haphazard nest and lays eggs. A supply of very effective venom may give the widow the confidence she needs. Unlike the male, the widow's fangs are large enough to penetrate human skin.

Females inject venom into an animal for two reasons—to protect themselves from harm and to immobilize the animal before eating it. Although a female is aggressive during prey capture, usually she must be pestered before using venom to protect herself. Usually this happens when she is threatened, for example, if a person handles a rag, tool handle, or log occupied by a black widow. Sometimes black widow spiders get on clothing or bedding and are squeezed when these materials are contacted. Also, spiders frequent the undersides of seats in outdoor privies. In all the aforementioned areas, humans invade the privacy of the spider. The result is a victim who experiences the effects of the widow's venom.

This venom causes a great deal of pain and discomfort. The site of the bite shows no obvious redness and resembles an area pricked by a needle, but symptoms are more obvious. Symptoms begin in about 10 to 20 minutes following the bite, and they include abdominal pain, vomiting, restlessness, sweating, elevated blood pressure, salivation, rashes, and convulsions.

Fortunately, treatment is available. Injection of an antivenom brings forth permanent relief in about one-half hour. Injections of calcium gluconate are also given to relieve muscle pain and cramps. Among those bitten, elderly individuals and children seem to react most severely to the venom. Therefore, they must be treated quickly and monitored frequently. However, all individuals, regardless of age, should seek medical attention immediately following a bite. Data from 1700-1943 indicate that the fatality rate from black widow spider bites was five percent. With modern techniques of treatment and the use of antivenom, this percentage will likely decrease.

BIBLIOGRAPHY

Anonymous. November 1995. Of sex, somersaults, and death. Discover: 34.

Dodge, N. N. 1952. Poisonous dwellers of the desert. 12th ed., revised. AZ: Southern Monuments Association.

Gertsch, W. J. 1949. American spiders. New York: Van Nostrand.

Rauber, A., M.D. 1983-1984. Black widow spider bites. Journal of Toxicology-Clinical Toxicology.

WATER CONSERVATION
IN CAMELS

Some individuals still believe camels store water in their hump or in a special bag associated with their stomach. This belief has been put in the category of folklore by researchers who have studied the camel on the Sahara Desert. Researchers recognize that water is definitely a scarce commodity for camels, but these researchers are convinced that camels carry no special reservoirs for water storage. They indicate that the camel's problem with water is twofold. One is to conserve water available to them, and the other is to function effectively with low amounts of water in their bodies.

Summertime, when shade temperatures soar to 120 degrees Fahrenheit, is when the camel's body displays the greatest effort to conserve and tolerate low amounts of water. The camel does not do this by taking a dip in a nearby pond, seeking shade, or by burrowing in moist soil. More sophisticated methods are used. First of all, camels excrete a minimal amount of urine—about a quart a day, and they do not sweat until their body temperature reaches 105 degrees Fahrenheit. Even on the hottest days, this body temperature is reached only for a short period of time. The overall idea is to keep body temperatures below 105 degrees to reduce water loss. This effort is accomplished by the camel's ability to lose heat through the skin. Heat loss is possible because body fat is stored in the hump and not beneath the skin as it is in many other mammals. Thus, in camels the flow of heat from the body outward is not restricted by a layer of fat. Also, a thick coat of hair restricts hot sun rays from penetrating the camel's body. Despite all these measures, the camel still ends up with low amounts of body water, but it has means to handle this problem. These means are best explained by comparing human body functions with those of a camel. In humans, blood thickens when it loses water. This overworks the heart and reduces blood output. With blood circulation reduced, metabolic heat of the body is not lost to the atmosphere from blood in vessels near the surface of the body. This increase in internal body temperature destroys enzymes which control metabolism, and the end result is death. In comparison, water loss does not occur in the blood of camels. Loss is from other body tissues and fluids. Therefore, the blood of the camel does not thicken as it does in humans. Blood continues to flow smoothly through the camel's vessel system.

Camels subjected to a great deal of water loss resemble walking skeletons with a coat of loose hide, but the hide smooths out quickly as the camel consumes water. Camels drink only enough to replace water which was lost, as much as 27 gallons in 10 minutes. And, guess what, they don't drink excess water for storage in the hump or a special bag associated with the stomach.

BIBLIOGRAPHY

Schmidt-Nielsen, K. 1968. The physiology of the camel, vertebrate adaptations: readings from the Scientific American. San Francisco: W. H. Freeman and Company.

COMMON VAMPIRE BATS

Believe it or not, vampire bats do exist. There are three species of vampire bats which inhabit South America and Mexico. All three species are about the same size, three inches from the tip of their nose to the base of their rump, a wing span of six or seven inches, and a weight of approximately one ounce. Vampire bats resemble other species of bats except in their diet.

Blood is the only food consumed by vampire bats. Two species feed on blood of birds. The third species, the common vampire bat discussed here, feeds on the blood of livestock, such as cows, horses, and mules. Vampire bats prefer the blood of livestock to blood of wildlife perhaps because of the accessibility of livestock. But a word of caution, humans who sleep with their bare skin exposed are not excluded from encounters with vampire bats.

Regardless of the source of a blood meal, the agenda of common vampire bats is the same. These bats depart once each night from roosts in caves, old wells, mine shafts, tunnels, hollow trees, or buildings and fly to feeding sights. Once there, the bats land on the ground and walk, run, or hop a short distance to the host. The bats climb the foreleg of the host, usually to the neck or shoulders, where blood vessels are numerous and close to the skin. Once common vampire bats find a suitable area for feeding on livestock, they use their razor-sharp, blade-like canines and incisors to make a cut in the skin several millimeters deep. Then the bats lap blood with their tongues as blood oozes from the wounds. Clotting is prevented by a chemical substance in the bat's saliva. Apparently, blood loss from the host is not enough to cause injury. Harm to the host results from a transfer of parasites, as well as rabies and other diseases from bat to host during the blood meal. For this reason, some methods of control are necessary.

Two methods are currently in use. One involves the injection of an anticoagulant into the stomachs of potential livestock hosts. This substance, which prevents blood from clotting, enters the blood circulation of the host. The anticoagulant is harmless to livestock when used in recommended dosages. Once the anticoagulant circulates in the blood of livestock, the anticoagulant becomes accessible to bats during their blood meal. Once ingested, the anticoagulant causes internal bleeding and death to the bat, possibly because of its high concentration in the small body of the bat. The other control method involves rubbing an anticoagulant on the fur of captured vampire bats. After returning to roost, this anticoagulant is spread from treated to untreated bats. The result is the same—internal bleeding and death. One obvious problem with this latter method is that harmless bats are also killed by the anticoagulant. Therefore, it is understandable why bat conservationists prefer injection of an anticoagulant into hosts more than any other method of control.

BIBLIOGRAPHY

Turner, D. C. 1975. The vampire bat: a field study in behavior and ecology. Baltimore: The Johns Hopkins University Press.

Vaughan, T. A. 1986. Mammalogy. 3rd ed. Philadelphia: W. B. Saunders Company.

AFRICAN LION,
KING OF BEASTS

The male African lion is honored with the title "king of beasts," a title unlikely to be held by any other animal. What is so special about the male African lion that gives it the right to a crown of royalty?

How about size? Males, the larger of the lion sexes, have a head and body length up to six and one-half feet and a weight of 550 pounds. Male African lions are definitely large cats, but the Siberian tiger is larger. It has a head and body length about two and one-half feet longer than the African lion and a weight 125 pounds heavier.

How about running speed? African lions can run 37 miles per hour for distances of about 100 meters, which is slightly greater than the length of a football field, but then again, the cheetah can run faster. They can run 70 miles per hour for up to 500 meters which is five times farther than the lion.

Do African lions hold the title because they roar louder than any other kind of cat? The roar of a male African lion is so loud that it can be heard up to five miles. Can their roaring ability be a criterion for high rank in royalty? That is a possibility, but perhaps certain aspects of their social behavior may account for the title "king of beasts."

African lions occupy social units called prides. A pride is composed of one to three breeding males, several breeding females, and their cubs. Males hunt infrequently and rest a great deal within the pride, but they have two roles which they take very seriously. One is to mate frequently with breeding females, and the other is to patrol the periphery of the pride's territory to guard against sexual encounters between females within the pride and males outside the pride. The sole function of males within the pride is to ensure that they are the only ones who inseminate the pride's females. Thus, only their genes, not those of outside males, transmit to the next generation of cubs. As long as this function is attained, males maintain a genetic governance. At that time, they may be referred to as "king of beasts." Such a role requires a tremendous amount of energy as well as a continuous state of alertness. Only a few mistakes from carelessness, poor health, or old age, and the monarchs of the pride are eliminated quickly by more alert, aggressive males from outside the pride.

BIBLIOGRAPHY

Bertram, B. 1978. Pride of lions. New York: Charles Scribner's Sons.

Guggisberg, C. A. W. 1975. Wild cats of the world. New York: Taplinger Publishing Company.

Nowak, R. M. 1991. Walker's mammals of the world. 5th ed. Vol. 2. Baltimore: The Johns Hopkins University Press.

Pusey, A., and C. Packer. August 1983. Once and future kings: groups of male lions compete for the chance to rule a pride and mate with its females. Their tenure is never secure. Natural History 92:55-57,60-62.

Schaller, G. B. 1973. Golden shadows, flying hooves. New York: Alfred A. Knopf, Inc.

UNIQUE FEATURES
OF ELEPHANTS

Several species of elephants lived in the past, but only two survive today. They include the Asiatic elephant and the African elephant. As their names imply, each inhabits a different continent—the Asiatic elephant, India and southern Asia; the African elephant, Africa south of the Sahara Desert. In spite of this, both live in similar habitats, namely, grassland, savanna, and jungle, but they have different diets. The Asiatic species prefers grasses and leaves; the African species, bark, branches, and roots. In addition to diet, many other features distinguish the two species.

The African elephant is the larger of the two. It can be one foot longer, three feet taller, and three tons heavier than the Asiatic. One outstanding difference is the size of the African elephant's ears. They are about three times larger than the Asiatic's. Also, African elephant tusks can be twice as large as tusks of the Asiatic elephant.

Large tusks are the main reason why African elephants are in great demand by commercial hunters. Tusks are carved into jewelry and other products and sold at high prices worldwide. Tusks are said to be composed of ivory, but ivory is really dentine because the tusk is a tooth, actually an incisor. A tooth consists of soft dentine on the inside and a layer of hard enamel on the outside. The enamel layer wears away early in elephants; thus, their tusk is composed entirely of dentine. Because of the small size of its tusk, the Asiatic elephant is not in demand for its ivory. But, there are other reasons why humans take Asiatic elephants from the wild.

The general consensus is that the Asiatic elephants are more docile than the African species. They are in demand as beasts of burden and as performers in circuses, but African elephants cannot be excluded entirely as performers either. After all, Jumbo was an African elephant. P. T. Barnum obtained Jumbo in 1882 from the London Zoo and made him the star attraction of the circus. Jumbo may be evidence that the African elephant is not as antisocial towards humans as most people would think. Perhaps both species display some evidence of being docile.

Other features are held in common by both species. Females have nipples between their front legs, unlike other kinds of grazing and browsing mammals which have nipples between the hind legs. Also, male elephants retain their testes inside the body. These testes are not carried in a scrotum like most other mammals. Both species perceive and transmit sounds in such low frequencies that they are not audible to the ears of most other mammals, including humans. These vocal transmissions may be useful to communicate with youngsters and other elephant herds. Males may also use these transmissions to avoid each other.

A discussion about elephants usually brings forth these questions. Do elephants fear mice and do they travel to an elephant graveyard to die? As far as authorities are able to determine, the answer is "no" to both of these questions.

BIBLIOGRAPHY

Allman, W. F., and J. Schrof. October 2, 1989. Can they be saved? (Kenya's effort to conserve its elephant population). U. S. News and World Report 107(13):52-58.

Contreras, J. November 18, 1991. The killing fields: officials in southern Africa say they must shoot elephants to protect them. Newsweek 118(21):86-88.

Nowak, R. M. 1991. Walker's mammals of the world. 5th ed. Vol. 2. Baltimore: The Johns Hopkins University Press.

Voelker, W. 1986. Natural history of living mammals. Medford, NJ: Plexus Publishing, Inc.

Tony Gennaro
'92

KILLER BEES

The name, African killer bee, definitely has threatening implications. However, toxin of killer bees has the same potency as the toxin of introduced European bees, more commonly known as honeybees, which have inhabited the Americas for several years. Both African killer bees and European bees are similar in appearance. One difference between these two species is that the killer bee is a recent arrival to the Americas.

The story behind African killer bees begins in Brazil in the 1950's. Brazil's honey industry had become dissatisfied with the productivity of honey from European bees. Therefore, the Brazilian ministry requested assistance from Warwick Kerr, a university professor, to improve beekeeping in Brazil. He imported African bees and began breeding experiments. In 1957, 26 African bee queens escaped from his experiments. They bred with European bees, producing Africanized hybrid offspring referred to as the Africanized bees. In 1964, military forces took over the government of Brazil. Kerr was critical of this government. To discredit Kerr, the military blamed him for all insect stinging incidents by stating that these incidents resulted from bees which escaped from Kerr's experiments. Military personnel named the escaped bees "killer bees". Brazilians soon became aware of Africanized bees, as well as people from countries outside Brazil where these bees were invading.

Africanized bees moved northward across Central America, and they have now entered North America. They travel about 200 miles per year, and they breed with European bees along their way. All people in their path are concerned, and they have good reason.

Africanized bees differ from European bees in several ways. When compared to the European bees, Africanized bees lay more eggs per queen, swarm more often, develop into adults quicker, begin pollinating earlier in the morning, continue to pollinate later in the evening, produce more honey per bee, are mite resistant, and are more disease resistant. Because of their nasty dispositions, they tend to be disturbed easier, will pursue in swarms rather than as individuals, will pursue for much greater distances, and are more apt to sting than the European bee.

It is obvious that only experienced beekeepers should handle Africanized bees. Inexperienced individuals who suspect the whereabouts of Africanized bees should keep their distance. The aggressive behavior of these bees could cause serious injury and even death. That is the bad side.

On the good side, the behavior of killer bees could be an important attribute. Authorities indicate that the arrival of the Africanized honeybee may eventually be one of the best things that has happened to American beekeeping. Africanized bees are relatively disease resistant and are an improvement of resident European bee stocks. After all, Brazil ranked 47th in the world for honey production before Kerr introduced African bees. In one generation with Africanized bees, Brazil now ranks seventh.

BIBLIOGRAPHY

Anonymous. August 12, 1985. Killer bees--the threat is for real. U.S. News and World Report 99:12.

Anonymous. August 12, 1985. Tracking an ill-tempered invader: scientists hunt for killer bees in California. Time 126:21.

Anonymous. August 1987. Killer bees advancing on U. S. USA Today 116:12.

Graham, F., Jr. March 1991. Killer (?) bees (threat of Africanized honey bee overestimated, says beekeeper). Audubon 93:14.

© Tony Gennaro 94

HUNTING ADAPTATIONS
OF OWLS

Owls are very distinctive in appearance and behavior. When perched, their erect, bulky-appearing bodies are quite unique. Except for occasional movements of the head from left to right, owls are quite rigid on the perch. Their eyes are conspicuously large, and their voices are typically low and monotone. Colors of owls are usually light and dark browns, except for the white of the snowy owl. All the variations of these few colors help the owl blend with the environment. There is a reason for this precise match of colors. Owls are predators, and it is important for them to be concealed from their prey which includes members of the rodent group, especially mice. Besides their ability to blend with the environment, owls have several rodent-catching features.

To begin with, feathers of owls are velvety in texture. This feature muffles the sound of flight; therefore, they can swoop down on a mouse in the dark without making the slightest noise.

Owls can see in the dark. This is possible because light sensitive cells in their retinas, called rods, are receptive to dim light. This feature distinguishes owls from day-dwelling birds which have light sensitive cells, called cones, in their retinas. Cones are receptive to bright light and color.

The eyes of owls are situated on the front of their head. This enables them to have overlapping fields of vision, meaning that owls can see an object in space with the left and right eyes at the same time. This gives them very good depth perception.

The sensitive ears of an owl surpass those of many vertebrates. If a mouse makes the slightest noise in the dark, this noise serves as a locator for the owl. The owl pinpoints the source of this sound on a horizontal plane which is parallel to the ground. It does this by rotating the head back and forth from left to right. Once locked in on that plane, the owl locates the sound on the vertical plane perpendicular to the ground. When the sound is located on the horizontal and vertical planes at the same time, the mouse is on target. The owl then swoops down silently in the dark and grabs the mouse with its talons. Evidence for the effectiveness of this audio lock-in mechanism has been shown for the barn owl. It can locate the sounds of prey on the horizontal and vertical plane in darkness with a deviation of only one degree. And, that is truly an accurate display of mouse-catching in the dark.

BIBLIOGRAPHY

Welty, J. C. 1988. The life of birds. 4th ed. Philadelphia: W. B. Saunders Company.

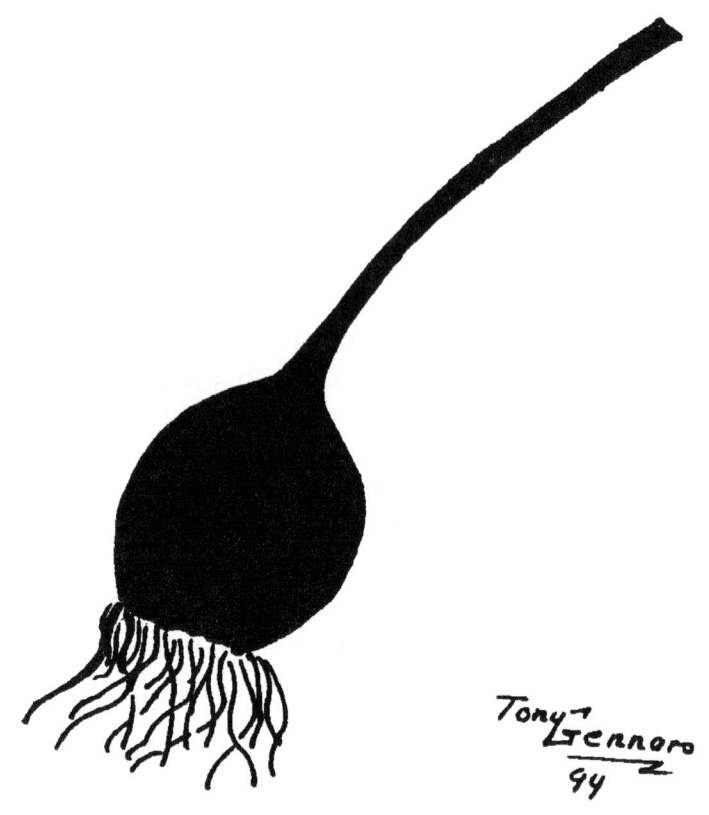

Tony Gennaro
94

MEDICINAL VALUE OF
GARLIC

Most of us are familiar with garlic, known primarily for its use in flavoring foods. Some of us cannot imagine certain foods without it, especially Italian meatballs.

Garlic has been used for many purposes other than flavoring foods. Some uses are physical, psychological, and spiritual. For example during medieval times, garlic was worn around the neck to provide protection from plague, vampires, werewolves, and demons. Garlic was fed to slaves who built pyramids to maintain their physical and moral strength. Alexander the Great fed this plant to his troops because he believed it increased their courage in battle. Even today, bullfighters wear garlic around their neck as good luck against horns of the bull. Garlic's use for medicinal purposes goes back in time, but research on its medicinal value continues today. Old, unproven remedies recommend garlic as a laxative and as a cure for snakebites and hemorrhoids. Recent uses of garlic stem from statements made by personnel of national cancer institutes, who declared that a diet rich in garlic can help prevent stomach cancer. Other studies have revealed garlic as an antidote for high cholesterol and high blood pressure, as well as an effective killer of bacteria, fungi, and yeasts. Also, researchers comment that it reduces the clotting capabilities of blood which is very useful in preventing heart attacks and strokes. Apparently, the chemicals in garlic that perform these healing feats are the smelly sulfur compounds. Unfortunately, they also give garlic its antisocial qualities. At the present time, these sulfur compounds are being isolated to determine exactly what medicinal value they have.

As far as human consumption is concerned, the big question is, "In what form should garlic be ingested—whole, crushed, cooked, or uncooked—to retain the effect of the sulfur compound?" Deborah Hufford in her article entitled, "A Bulb of Benevolence," indicated that beneficial chemicals are obtained from garlic when it is heated or cooked. Therefore, either method is recommended. However, the bulb's healthful aspects may be destroyed if it is deodorized or prepared as a powder or garlic salt.

No one has commented on the quantities of garlic necessary for cures and ailments regardless of the manner in which it is consumed. Some authorities say that large amounts are harmless. Others have reported cases of allergies, stomach disorders, and diarrhea among heavy users. We need to be patient with this interesting plant because further research is necessary to unlock its medical secrets. Probably the best thing we can do at the present time is heed Deborah Hufford's suggestion to change the old saying, "an apple a day keeps the doctor away."

BIBLIOGRAPHY

Hufford, D. May-June 1990. A bulb of benevolence. Saturday Evening Post 262:26-28.

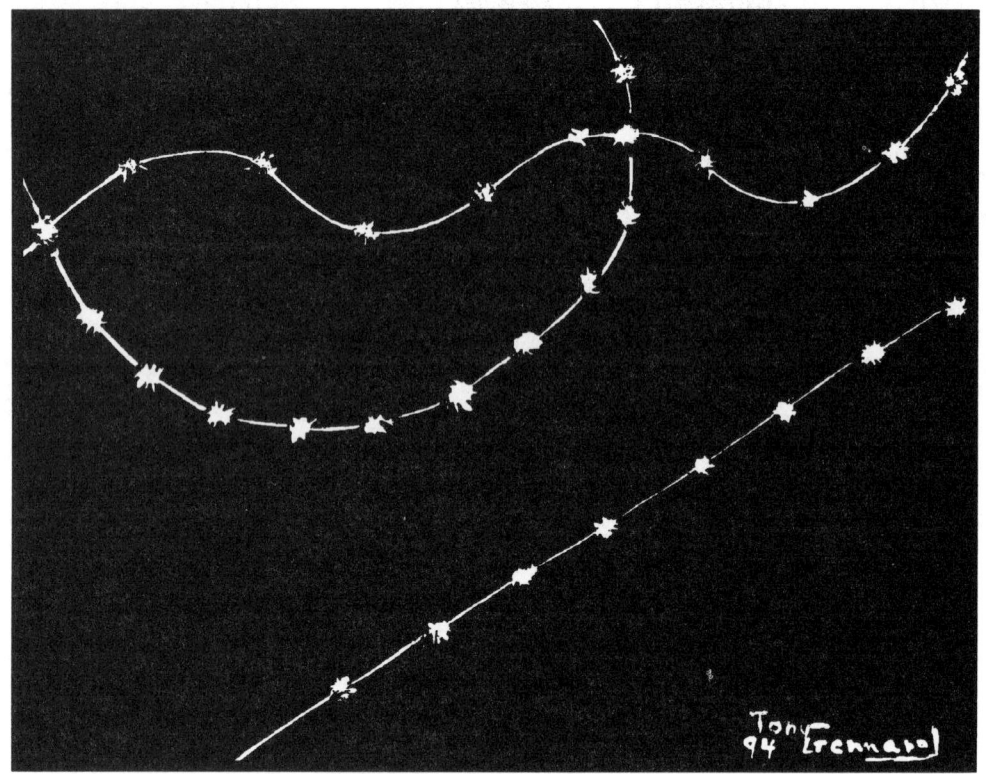

FIREFLY BEHAVIOR

Most of us are familiar with fireflies or lightning bugs, which are really beetles, not flies or bugs. The tiny flashes of light they emit certainly attract one's attention on a warm summer evening. While observing them, questions come to mind, such as what causes fireflies to flash, and why do they flash?

The tiny specks of light we see flashing and streaking in the darkness result from chemical reactions in cells on the underside of the firefly's abdomen. These flashes and streaks are red, orange, or green, depending on the species. Only adults flash, but even the worm-shaped larval stage of the firefly glows.

Fireflies flash to allow males and females of the same species to recognize each other in the dark because flashes of fireflies are distinct to species. When recognition is fully accomplished by both sexes, they court and mate. The overall procedure is like this. The female emerges from her burrow on the ground as darkness approaches, and she remains on the ground or perched in a bush. A male in search of a female displays his flash patterns. If he happens to fly near her, and she recognizes these patterns as being the same as those displayed by her species, she answers with her flashes. They continue to flash and streak back and forth to ensure recognition. Finally, he lands near her. They copulate and afterwards, she returns to her burrow. He then becomes airborne again and initiates a flash pattern as he searches for another female.

This unusual mating procedure seems benign enough, but it is not always peaceful. Females of some species of fireflies are carnivorous, meaning they eat males of other species. These females have broken the code of flash and streak patterns between sexes of other species. When a male makes a fly-by with flashers on, the carnivorous female on the ground or in a bush responds by flashing back a pattern that is distinctive to his species. She knows she is not dealing with a male of her species, and her intent is certainly not courtship. They continue to exchange flashes, and finally the male lands next to her, and to his surprise, he becomes her tasty meal.

Fortunately, that is not the end of the story. Not all males called in by these deceiving females are embraced and ingested. Some of them are lucky enough to escape and learn from the experience. The next time such fortunate males observe familiar flash patterns from below, they land a distance away from the female and check out the situation. Those perceptive enough to detect the intentions of a lurking, hungry carnivorous female, fly away and eagerly await another night to signal a female of their own species.

BIBLIOGRAPHY

Limonick, M. D. November 17, 1986. Of fireflies and tobacco plants. Time 128:87.

Lloyd, J. E. July 1981. Mimicry in sexual signals of fireflies. Scientific American 245:138-145.

Peterson, I. August 31, 1991. Step in time: exploring the mathematics of synchronously flashing fireflies. Science News 140:136-138.

THE MALE AFRICAN LION

The life of a male African lion begins in a dense patch of vegetation, well-hidden from hyenas and other predators. This birthplace is usually a few miles from the site where an extended family is located. This family is a social unit of lions called the pride. About six weeks after birth, the young male's mother leads him along with brothers and sisters out of seclusion to join the pride. Members of the pride include the young male's father, one or two uncles, and several females which are his mother's sisters, half sisters, and cousins. Cubs born to those females are active members of the pride. At their earliest ages, all the cubs behave alike. As they grow and sexual hormones become functional, appearances and behaviors of males deviate from those displayed by females of about the same age. As three-year-old subadults, males leave the pride voluntarily or are forced out by adults.

Two or three males, usually brothers, wander together as nomads for at least three years. Since food is difficult to find, often these male lions survive by eating scraps of meat remaining from the kills of hyenas or other lions.

When these brothers are mature enough, they chase off the resident males of an existing pride. Soon, breeding females accept the invading males socially, and the new males assume leadership of the pride. Males copulate with breeding females and father the cubs. Since invading males are unrelated to the females, genetic variability of cubs is achieved.

Invading males display no tolerance of cubs fathered by previous males. This behavior may exist because there is no benefit for male lions to invest time with cubs which contain genes of another male. Therefore, current governing males usually kill cubs fathered by previous male residents of the pride. Following the death of the cubs, their nursing mothers cease milk production and become receptive to the resident males. Eight months later a baby boom occurs.

Once paternity is established, the life of males becomes routine. They mate frequently with receptive lionesses, eat prey killed by lionesses, guard the edges of the pride's territory against intrusion by nonresident males, and sleep 20 hours out of each 24-hour day.

These adult males have a pride tenure of about three years. After this time, aging has taken its toll, and males lose the vigor they once displayed. During earlier years, breeding males mostly associate with their fellow brother governors, not lionesses, except during copulation and feeding time. As these males age, they begin to associate more often with the lionesses.

Aging males are soon driven from the pride by a coalition of young aggressive males eager to mate with receptive females of the pride. The expelled males become wandering nomads. Because of their age, typically about ten years, they find food with more difficulty than when they were younger. Life is hard for these lone, ejected hunters, but fortunately it is short. They typically die as aged nomads in a year or two.

BIBLIOGRAPHY

Bertram, B. 1978. Pride of lions. New York: Charles Scribner's Sons.

Guggisberg, C. A. W. 1975. Wild cats of the world. New York: Taplinger Publishing Company.

Nowak, R. M. 1991. Walker's mammals of the world. 5th ed. Vol. 2. Baltimore: The Johns Hopkins University Press.

Pusey, A., and C. Packer. August 1983. Once and future kings: groups of male lions compete for the chance to rule a pride and mate with its females. Their tenure is never secure. Natural History 92:55-57,60-62.

Schaller, G. B. 1973. Golden shadows, flying hooves. New York: Alfred A. Knopf, Inc.

SPOTTED HYENAS

Would you be proud to walk a hyena through a crowded park? Probably not, because they are stereotyped as being sneaky creatures that roam around at night eating scraps of meat, bone, and hide from prey killed by other animals. Also, their normal body position, high on the forelimbs and low on the hind, signals a stance of submission to humans—sneaky submission, that is.

Surprisingly, not all species of hyenas are solely scavengers. Some kill prey. Of the four species of hyenas which inhabit Africa and Asia, the aardwolf preys on insects, the brown hyena is primarily a scavenger, the striped hyena hunts in some areas and scavenges in others, and the spotted hyena hunts most of the time. In fact, most dead prey fought over by spotted hyenas and African lions are actually killed by the hyenas, not the lions.

Spotted hyenas live in well-organized social groups or clans. A clan of up to 80 individuals occupies an area as large as 12 square miles. The nursery, where mothers suckle their pups, is situated in the middle of the clan's territory. All spotted hyenas of the same clan recognize each other, and the area they occupy is defended and marked off with their anal gland secretions.

Clan members hunt in packs. Small packs capture and kill wildebeests, and large packs kill zebras. In all cases, hyenas are ferocious eaters. Their jaws are more powerful, and their digestive capabilities are greater than other carnivores. Hyenas digest flesh, bones, and hides very readily. This aggressive digestion is accomplished by a high concentration of acids in their digestive tracts.

Spotted hyenas are unique carnivores in another way. They display a reversed social rank, in that females are dominant over males in clan activities. Females are the first to obtain primary resources such as food and other substances essential for survival. This female dominance is brought about by high levels of the hormone testosterone.

Physically, the genitalia of females have the same appearance as that of males. However, this is only for appearances in order to gain social rank because females remain functional in their own sex. They produce offspring and suckle their pups like females among other species of carnivores.

Spotted hyenas, commonly referred to as "laughing hyenas," are not laughing at all. They "laugh" only when attacked on chased.

BIBLIOGRAPHY

Nowak, R. M. 1991. Walker's mammals of the world. 5th ed. Vol. 2. Baltimore: The Johns Hopkins University Press.

ENERGY FLOW IN ECOSYSTEMS

Energy is defined as the ability to work. In turn, work among living organisms is expressed as their functions. Functions are eating, running, sleeping, digesting, reproducing—in other words, living. This energy for living is obtained from ecosystems.

An ecosystem (pronounced with a long e) is a distinct environmental unit, ranging in size from a tiny rotting log to the entire planet Earth. An ecosystem could be a stand of evergreens in a forest, a stand of grasses on the prairie, a town, city, state, the United States, North America, or all the Americas. The two major parts of the ecosystem are the nonliving and living components. Nonliving components include air, soil, rocks, and water. Living components include plants, animals, and microorganisms. These living components are ranked in the ecosystem on the basis of their proximity to the initial source of energy—the sun. The first level consists of plants, called producers, because they are the original sources of energy for the system. The energy they produce in their tissues is derived from the radiant energy of sunlight. This energy is passed on to animals of the second level, called primary consumers, when they consume plant tissues. A few examples of primary consumers are grasshoppers, rodents, birds, cattle, horses, elk, sheep, and goats. Secondary consumers are in the next level. They obtain energy by ingesting primary consumers. Examples of secondary consumers are insect-eating birds, flesh-eating rodents, snakes, raptors, and lions. Some of these secondary consumers are eaten by tertiary consumers, such as coyotes and other carnivores. Various kinds of animals may occupy more than one level. For example, coyotes can be primary consumers when they consume berries, secondary when they consume rodents, and tertiary consumers when they ingest snakes. Likewise, humans are primary consumers when they ingest plants and secondary consumers when they eat meat from livestock. They can also be tertiary consumers when they ingest food such as shark steak. And finally, decomposing microorganisms consume nonliving tissues from all levels of the ecosystem.

The efficiency of energy captured in this system, and the efficiency of its transfer through the system is a concern to human societies. To begin with, not all energy of sunlight is available to plants. Only a small part is fixed in plant tissues because most of the sun's energy is reflected as heat or lost in other ways. In turn, only a small part of the energy within plant tissues is available to primary consumers because it is used for plant metabolism or lost as heat to the atmosphere. Again, energy used for metabolism and lost as heat among primary consumers is not available to secondary consumers. The same is true for subsequent consumers.

This energy loss is of considerable importance when the problem of increasing human populations comes to mind. Two technological efforts are in progress. One is to decrease energy loss during its original capture by plants, and the other is to increase the efficiency of energy transfer from one level to the next. The purpose is to increase the availability of food energy to the level consumed by humans. In turn, this increase will bring about a greater production of human tissue. In simple terms, it will provide increasing amounts of food energy to satisfy the demands of increasing human life.

BIBLIOGRAPHY

Starr, C., and R. Taggart. 1992. Biology, the unity and diversity of life. 6th ed. Belmont, CA: Wadsworth Publishing Company.

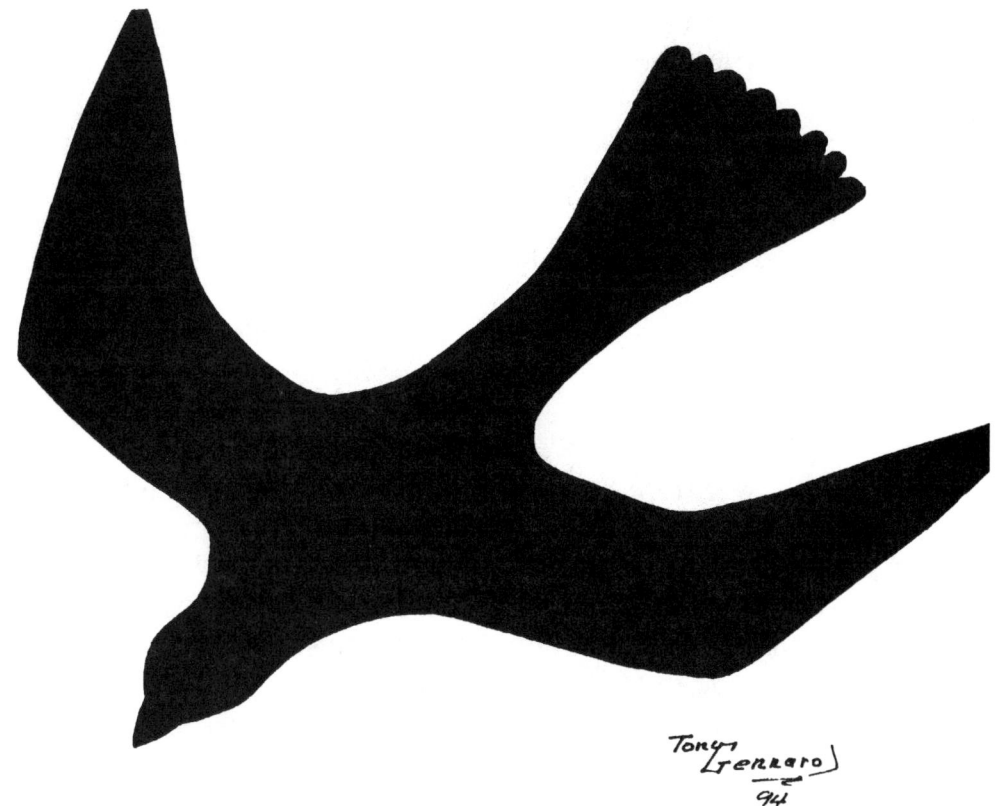

Tony Ferraro
94

HUMAN CONFLICT WITH WILDLIFE

Wildlife was an important resource to early humans, and therefore, the two rarely competed with each other. However, as agriculture and domestication became part of human societies, competition with wild animals for resources began. In the aggressive struggles that followed, humans were rarely the loser, as is the case today. Unfortunately, current confrontations are not restricted to issues concerning livelihood. Confrontations extend into recreational activities—golf, for example.

As a research biologist, I dealt with an interaction between golfers and wildlife. The wildlife involved was a bird called the Mississippi kite. These grayish falcon-like birds have a 14-inch body and a 3-foot wing span. Kites winter in central and northern South America and summer in southern United States. Within their summer range, kites typically inhabit rural areas in the Southeast and urban areas in the Southwest.

In the latter area, golf courses are a favorite urban nesting site. These areas are preferred by kites because trees are generally spaced far enough apart to allow kites room to forage for flying insects. Usually, kites on golf courses conduct life functions unnoticed, at least until they dive upon golfers who approach kite nests. Then, golfers react to this startling experience by uttering profanities and swinging their clubs wildly at diving kites. Such behavior led me to search for a less violent remedy to eliminate conflict between these two opponents.

Through my research, I discovered some interesting facts about kites. I noticed that mate selection was complete at the time kites arrived on the breeding grounds. Nests built the previous year were used or new ones were built. I never observed more than one kite nest per tree, and kites never constructed a nest in a tree which was closer than 25 yards to another tree with a nest.

Other observations were noted during my research. Kites performed several activities on the breeding grounds, including nest site selection, nest construction, egg incubation, and rearing of nestlings. Kites budget time effectively to conduct these functions, and they performed them in sequence.

With this sequence in mind, as well as the commitment of those functions to time, I developed a plan to outsmart the kites. My guess was that eight days or so was not enough time for kites to habituate to alarm signals in the environment. For example, it probably takes more than seven days for other birds to habituate to a scarecrow. Therefore, I concluded that eight days was not enough time for kites to distinguish between life-sized kite models and real kites. I then placed life-sized kite models in trees before kites arrived on the breeding grounds. Kite models were placed in nests built the previous year, or in artificial nests placed in trees which lacked nests. As kites arrived, they recognized the models as their own kind and built their nests farther than 25 yards from trees with models. This technique kept kites away from areas used by golfers.

This research on Mississippi kites proves it is possible to reduce or eliminate human and wildlife conflicts without destroying wildlife. In the case with Mississippi kites, knowledge of species behavior provided clues to solve the problem at hand.

BIBLIOGRAPHY

Butelho, E. S., and A. L. Gennaro. 1993. Parental care, nestling behaviors and nestling interactions in a Mississippi kite (*Ictinia mississippiensis*) nest. J. Raptor Res. 27(1):16-20.

Gennaro, A. L. 1988. Extent and control of aggressive behavior toward humans by Mississippi kites. Pages 249-252 *in* R. L. Glinski et al., ed. Proceedings of the Southwest Raptor Management Symposium and Workshop. Natl. Wildl. Fed., Washington, D. C.

Glinski, R. L., and A. L. Gennaro. 1988. Mississippi kite. Pages 54-56 *in* R. L. Glinski et al., eds. Proceedings of the Southwest Raptor Management Symposium and Workshop. Natl. Wildl. Fed., Washington, D. C.

FLEAS

A flea on a patch of bare skin covering a rich supply of blood vessels is as happy as a landowner standing on top of a rich supply of oil. There is good reason for the flea's joy. Blood is its sole food resource. Fleas obtain blood by piercing with their beaks the skin of their host. Then, they suck the blood out of vessels with specialized mouthparts. In addition to being able to obtain a meal of blood, fleas remain on a host that moves about. Also, they resist reactions from hosts irritated by their presence.

Certain traits are required for such unique adaptations. For one, clinging and moving is possible because of three pairs of strong, spiny legs. In addition, the bodies of fleas are flat from side to side. This enables them to pass between the bases of feathers and hairs. Because of their hard skin, fleas are resistant to the host's scratches and bites. Fleas are small in size, in some cases, as small as the letter "o" on the printed page. A small size reduces the chance of detection; therefore, fleas are well-adjusted for life on animals they parasitize. It is also beneficial for fleas to move from one host to another and to spend time in an area away from the host. The latter area is the host's residence.

Host departures serve numerous functions. For example, one host may provide a more abundant blood supply than another. Also, once a host dies, fleas must find another host because blood of the dead host cools and quickly decays. Egg-laying generally takes place in the residence of the host; therefore, interhost or off-host movements are essential. Here is how such activities are possible.

Fleas are great jumpers. Some can leap as high as seven and three-fourth inches and broad-jump thirteen inches. During a jump, they cartwheel through the air with front legs pointing upwards over their back, and the third pair of legs pointing downwards so the flea can grab the host regardless of what part of its body makes contact on landing. At other times, a cartwheeling flea may purposely land on a substrate where the host resides.

About 80% of a flea's life is spent in the residence of the host, which could be a rodent burrow, coyote den, or carpet of a house. Eggs are laid in those areas by adult fleas. Eggs hatch into larvae which are legless, worm-shaped, and whitish in color. Larvae have chewing mouthparts and consume debris and feces of the adults. Larvae change into pupa stages inside a cocoon and then emerge as adults within 28 to 42 days from the time eggs are laid, depending on the species. Therefore, as long as suitable conditions are available for fleas, their basic essentials of life are met.

On the other hand, from the standpoint of disease transmissions from fleas to humans, to domestic animals, or to wildlife, flea control is a necessity. This is accomplished by chemical applications to hosts and areas where they reside. Using infested pets as an example, treatment should be applied to them first, followed by treatment and vacuuming of carpets, throw rugs, upholstery, pillows, sofa cushions, couches, mattresses, or any place occupied by the pet. Be certain to consult a veterinarian about the nature of the chemical, do not over treat the pet, and discard the vacuum bag.

Flea control on wildlife species may occur without the assistance of humans. Lucy W. Clausen in her book entitled, "Insect Fact and Folk-lore," mentions a story from the Saturday Evening Post which deals with the subject of flea control in foxes. Ms. Clausen stated that a fox collected sheep wool caught on barbed wire. The fox placed the wool in its mouth and then trotted to a nearby pond. At the edge, the fox turned around and slowly backed into the water. The fox submerged itself except for the tip of its nose and the wool in its mouth. Then, the fox suddenly leaped from the water, dropped the wool, and ran away. Upon examination of the wool, it was found to be "alive" with fleas. This means that fleas cannot breathe underwater. Certainly this fox tactic will not be observed very often, if at all, but it does appear to be a good strategy.

BIBLIOGRAPHY

Askew, R. R. 1971. Parasitic insects. New York: American Elsevier Publishing Company.

Barker, W. 1960. Familiar insects of America. New York: Harper.

Borror, D. J. 1989. An introduction to the study of insects. 6th ed. Philadelphia: Saunders College Publishing.

Chapman, R. F. 1969. The insects: structure and function. New York: American Elsevier Publishing Company.

Clausen, L. W. 1954. Insect fact and folklore. New York: Macmillan Publishing Company, Inc.

Lanham, U. 1964. The insects. New York: Columbia University Press.

Tony Gennaro 94

CELLULOSE DIGESTION

Humans cannot digest cellulose. If they could, Huckleberry Finn would have eaten straws instead of chewing them. Cellulose is the fibrous component of wood and is that part of a straw which makes it stiff. Organisms which digest cellulose include certain kinds of bacteria. These bacteria use the chemical elements from cellulose to make up vitamins and proteins for their own bodies.

Cellulose-digesting bacteria reside in the digestive tract of herbivores. Herbivores live on a diet of plants. Bacteria benefit from herbivores because of the stable environment and supply of cellulose provided to them by the herbivore. At the same time, herbivores digest and obtain their nutrients from the bacteria.

Certain kinds of herbivores are called ruminants. These maintain cellulose-digesting bacteria in their digestive tract. Ruminants include cattle, sheep, goats, deer, elk, moose, and pronghorn antelope. Ruminants have four chambers, one next to the other, along the gut which are specialized to manage the functions of the bacteria. The first three chambers are in the esophagus which is the digestive tube between mouth and stomach. The fourth chamber, which follows the esophagus, is the stomach.

This multiple-chambered unit, or four-chambered stomach as it is commonly called, functions in the following manner. When vegetative material, called ingesta, enters the ruminant's mouth, saliva is added to the ingesta as it is ground into smaller particles by teeth. Once swallowed, ingesta enters the first two chambers, where it is churned by muscular action. Bacteria in those chambers begin digestion of cellulose to acquire proteins and vitamins for their own cellular functions. Then, some of the churned ingesta, which is not small enough in size, is regurgitated back to the mouth as cud. The cud is rechewed, swallowed, and reworked by bacteria again in the first chamber. It is then moved on to the third chamber where water is removed. Finally, all bacteria and other components of the ingesta pass into the stomach where cellulose-digesting bacteria meet their final destiny. They are digested, and the nutrients in their cells, which they prepared for themselves, are absorbed by the ruminant's intestine and used by the ruminant to sustain its own life.

Nonruminants, that is animals which consume plants and lack multiple chambers, usually contain cellulose-digesting bacteria in their large intestine and cecum. The latter is a small sac situated at the junction of the small and large intestines. Nonruminants include horses, asses, rhinos, and other herbivores.

Cellulose digestion is not as efficient in nonruminants as it is in ruminants. Fecal material of cattle, for example, is very pasty and lacks the undigested cellulose visible in horse fecal droppings.

Why should we ingest cellulose if we lack cellulose-digesting bacteria in our intestinal system? Cellulose, which is in wheat bran, whole-grain breads, cereals, and some vegetables, is a recommended part of our diet because it provides bulk, also known as fiber, or roughage, which is actually undigested cellulose. Fiber initiates bowel movements, promotes regularity, and as a result, it reduces the amount of time that cancer-causing agents in the feces have to act on tissues lining the colon. Therefore, cellulose is an important component in human diets.

BIBLIOGRAPHY

Church, D. C. 1971. Digestive physiology and nutrition of ruminants. Corvallis, OR: O & B Books.

Jurgens, M. 1982. Animal feeding and nutrition. Dubuque, IA: Kendall/ Hunt Publishing Co.

Mader, S. L. 1990. Biology. Dubuque, IA: Wm. C. Brown Publishers.

Moen, A. 1973. Wildlife ecology, an analytical approach. San Francisco: W. H. Freeman and Company.

POLAR BEARS

Polar bears are heavy, long-necked mammals with a coat of thick white fur. They are typically observed strolling along ice of the Arctic region alone or with cubs. Observations of mating pairs are uncommon because they pair only for a short period in early spring.

Pregnant females generally move inland 4 to 34 miles to hibernate and give birth in dens which they excavate in steep snowbanks. During hibernation, respiratory rates and body temperatures decrease, but bears do not enter a deep sleep. Females give birth between November and January to a litter ranging from one to four cubs, usually two. Each newborn weighs a little more than a pound and is about eight inches long. Following birth and departure from the den, females with young avoid interactions with adult males because they are potential predators of cubs. Cubs leave their mothers at about two years of age.

Inland areas are not a favorite habitat for polar bears. Both sexes prefer pack ice. This consists of large floating masses of ice fragments resealed to one another as a result of freezing temperatures.

Pack ice exists throughout the range of polar bears in the Arctic region, and bears occupy this ice year-round. However, inhabitants in the southern part of the polar bear range move to land during summer because pack ice melts.

Cape Churchill, along the shores of Hudson Bay in northeastern Manitoba, is one of the areas occupied by polar bears in summer. Since humans occur there as well, bear-human conflicts are unavoidable. These conflicts have demanded assistance from wildlife personnel who trap, tranquilize and confine the bears until they are released in early winter. Neither humans nor bears are happy that the other occupy Cape Churchill in summer.

Hardships for bears in Cape Churchill are primarily induced by the environment. Bears experience discomfort from heat and scarcity of food. They consume food items which are less than their first choice, such as birds, fish, eggs, vegetables, reindeer or caribou, and various stranded mammals on beaches. The favorite prey of polar bears, ringed seals, are not available because they remain on the pack ice.

Seal-hunting on pack ice has developed into a fine art for polar bears. Capture success has been enhanced by the bear's white coat which matches a background of snow and ice. This color allows the bear to sneak up on unsuspecting ringed seals which reside among the cracks and crevices of pack ice. At other times, a hungry bear will lie and wait near a hole in the ice from which a seal is likely to emerge. The bear can remain there for hours. Some naturalists have commented that the bear may be sleeping instead of waiting. Obviously, no one has gotten close enough to find out. Overall, this kind of stalking and waiting may imply that polar bears are slow and sluggish—not so.

Physical capabilities of polar bears are quite surprising. They can swim 4 miles per hour for distances of up to 40 miles. In fact, they can even outrun reindeer or caribou for short distances. As fast as they may be, polar bears rarely escape the capture and kill techniques of humans.

Polar bears are a valuable resource for humans. Native people of the Arctic region have hunted polar bears for years for their fur and fat. These bears are also in demand for sport and commercial hunting because today polar bear pelts are more valuable than the pelts of any other species of mammals in North American markets.

Recent reports indicated that the number of polar bears is stable and possibly increasing. This could be attributed to cooperative agreements by the United States, Canada, Denmark, Russia, and Norway which restrict hunting, protect habitat, and call for cooperative research on polar bears. However, the future still appears bleak for polar bears. Oil and gas exploration in the Arctic may likely threaten their numbers because these activities disturb inland denning sites.

BIBLIOGRAPHY

Davids, R. C. February 1985. Masters of the arctic ice. Science Digest 93:38-42.

Gup, T. September 16, 1991. In search of the great white bear. Time 138(11)70-71.

Nowak, R. M. 1991. Walker's mammals of the world. 5th ed. Vol. 2. Baltimore: The Johns Hopkins University Press.

Stirling, I. January 1989. Sleeping giants; sometimes the struggle for survival requires a lot of loafing. Natural History 98(1):35-39.

Wiley, J. P., Jr. March 1986. Polar bears mingle with people on Manitoba's Cape Churchill. Smithsonian 16:41-48,50.

Tony Gennaro
94

A HUNTER — THE
JUMPING SPIDER

Some species of spiders are web-builders and rely on capturing prey in their web, while others are stalkers. Stalkers do not build webs. Instead, they search for their prey. Jumping spiders are stalkers.

Jumping spiders are unique in appearance. They are covered with a hair-like material and grow to a length of one-half inch. In the southwestern United States, jumping spiders are typically flat-bodied, often black in color, and have a red or white spot on the top of their body behind the head area. These spiders have a large conspicuous pair of eyes on the median front part of their head which appears to be glaring back at anyone who is observing the spider. The jumping spider cannot be confused with the dangerous, web-building black widow spider. The black widow has a large, black, hairless, pea-sized abdomen, which has a bright red coloration on the underside.

Jumping spiders generally trail a single strand of silk behind them. They may suspend themselves from various objects with this strand, or they may be observed crawling around on windowsills or other areas while in search of food. Occasionally, they will jump from one place to another. This procedure is displayed primarily during prey capture.

Prey capture tactics of jumping spiders are quite predictable. Two pairs of eyes on the sides of their head detect motion. The one pair of large eyes located on the front of the head detects images in color. A smaller eye on each side of the large front eyes detects distance. All eyes see effectively up to 14 inches, but their best vision is at about 4 inches. With all eyes operational, these spiders move along slowly until they detect motion. They then turn their body in that direction and focus the large eyes on the prey, determine the distance of the prey, and then approach it slowly until they get close enough to jump on it. The spiders then jump, capture, and pierce the body of the prey with their fangs. Venom paralyzes victims and renders them motionless. Finally, jumping spiders extract juices from prey.

Although the venom of the jumping spider is poisonous, as is the venom of most spiders, it is not dangerous to humans. Thus, these spiders can be observed without fear of a lethal spider bite. Observations might reveal some unique features.

These spiders are extraordinary jumpers. Some can jump 40 times their length, a distance of about 20 inches. This is equivalent to a 6-foot human broad-jumping 240 feet, which is incredible, considering the human broad-jump record is approximately 29 feet. Jumping spiders, however, do not make these jumps without an element of safety.

The single strand of silk they trail from their bodies is a life-line. It is attached at one end to the spider, where it is produced, and the other end is attached to an anchoring object. If these spiders misjudge their jumps and fall from a leaf or branch, the life-line acts similar to a bungee cord, except that it does not stretch the same distance. Face it, these spiders have been bungee-jumping for over 300 million years, and humans discovered that thrill a few years ago.

BIBLIOGRAPHY

Preston-Mafham, R., and K. Preston-Mafham. 1984. Spiders of the world. New York: Facts on File Publications.

Sinclair, S. 1985. How animals see: other visions of our world. New York: Facts on File Publications.

Tony
Beckham
94

BIRD MIGRATION

Migration involves movement to a particular area and a return to the original place of departure. Migration is sometimes confused with emigration, a journey out of an area, and with immigration, a journey into an area. Birds are the most common kind of animals that migrate.

Migration enables birds to experience optimum conditions year-round. For example, most of the approximately 400 migrant species spend their summers in northern areas and their winters in southern areas. At the time they occupy either of these areas, weather conditions and food resources are favorable. The highest risk for a migratory way of life is during movements between summer and winter ranges.

The greatest hazard during migratory flights is inclement weather. Birds generally depart for their journey when weather is favorable; however, late spring and winter snow, rain, and windstorms cause high bird mortality.

Mortality risks are proportionate to distances of travel. Some species of hummingbirds and quail, which simply travel up mountain slopes for summer and down mountain slopes for winter, confront fewer hazards than long-distance birds such as Arctic terns which travel 24,000 miles round trip annually from their breeding grounds in the Arctic to their wintering grounds in the Antarctic. How is it possible for birds to travel these long distances?

The answer lies in the efficient storage and metabolism of body fat which yields twice as much energy per gram than proteins and carbohydrates. This fat is stored in special tissues under the skin, in muscles, and in the body cavity. In the white-crowned sparrow, for example, fat is deposited initially at 15 different sites and eventually forms a continuous layer between the skin and muscles. Having this energy source to make long flights is remarkable, but equally incredible are the inherent mechanisms which initiate migratory flights in the first place.

Increasing day length is correlated with northward migratory flights in the northern hemisphere. This is reasonable because increasing day length is associated with the onset of spring, a time of year when the best resources are available. Increasing light is perceived by the retina or through the skin of the bird. Nerves are stimulated, and these impulses are sent to the brain. The brain, in turn, starts hormone production in the blood which increases the size of testes and ovaries, increases body fat, and initiates an inherent urge to migrate.

Because members of the same species are so genetically similar, their departure times are highly synchronized. Thus, they arrive on the breeding grounds at approximately the same time. This arrival begins the countdown of an inherent internal time-budget which monitors tightly scheduled breeding events, such as pair formation, nest building, egg-laying, egg incubation, and rearing of the young before and after they leave the nest. All events must be completed before the time arrives for a return flight to the winter grounds.

Certain events are correlated with the readiness of birds to fly in the fall. They include maturity of the young, prior exposure to long days, numbers of days since the birds departed from their wintering grounds in the spring, decreasing day length, favorable weather conditions and other factors. Whatever the causative factor is, one thing is necessary, and that is winter migrants must arrive at their winter homes before inclement weather begins.

BIBLIOGRAPHY

Gill, F. B. 1995. Ornithology. 2nd ed. New York: W. H. Freeman and Company.

Plant Succession

Various kinds of vegetative patterns can be found on planet Earth. Examples include tropical rainforests near the equator, grasslands on treeless plains, and coniferous forests near polar regions. Any one of these patterns can be in balance, or equilibrium, with climate. In equilibrium, each pattern is referred to as a climax. A climax pattern will remain in place and persist as long as the area occupied by the climax is not disturbed, and if the region's climate does not change. Should disturbances or climatic changes occur, these climax patterns are temporarily replaced by other vegetative patterns referred to as subclimaxes.

Vegetative climaxes vary with latitude and altitude. For example, from the equator northward or from the equator southward, vegetative climaxes change latitudinally from tropical rainforest to grassland, woodland, coniferous forest, to tundra near the poles. Climax vegetation also changes with altitude. In southwestern U.S. mountains, for example, climaxes change from grassland to coniferous forest to tundra, as altitude increases. These changes occur because the environment gradually cools with increasing northward or increasing southward lati-

latitudes from the equator, and as well, with increasing altitudes.

It is unlikely that climax vegetative patterns could occupy all areas of the planet at the same time. This can occur only in absence of environmental disturbances wherever a climax exists, but disturbances are common. For example, glaciers removed all the vegetation and soils and exposed bare granite in northern parts of North America during the last glaciation. Other changes include floods, volcanism, moving dunes, earthquakes, landslides, and fires. Subclimax vegetative patterns, rather than climax patterns, follow these disturbances.

There are reasons why climax vegetation does not return immediately, such as substrate changes caused by the disturbance. Climax vegetation will not grow on bare rock, lava, or water. This example of climax plant species failing to return immediately to an area following a disturbance is experienced by anyone who cultivates a garden in a grassland area. Following the planting of seeds in soils disturbed by tilling, weeds germinate along with the garden seeds, not the original grass species. What accounts for the presence of these weeds—or forbs—as they are called botanically?

Certain specialized species of forbs are adapted to grow in disturbed areas. Therefore, they are part of the first successional stage, sub-climax stage, or sere, to occupy the disturbed area. Other seres follow the first sere. Plants from each sere improve the soil in many ways. They provide shade, decrease wind velocity, and decrease loss of moisture. Additionally, when plants die at the end of a growing season or from other causes, they add humus to the soil. Because plants do such a good job of soil improvement, species of one sere are replaced by more competitive species of a succeeding sere. Seres continue to replace each other in this manner until the climax returns.

A good example of a succession of plant seres involves pond succession in a climax forest. In this situation, a pond has a short life because it will eventually be replaced by the various species of the climax forest. The first sere of pond succession consists of algae which grows on the bottom of the pond in the shallowest parts around the edge. As some of the algae reach longevity, their decaying tissues add substrate to the bottom of the pool and decrease its depth. As the pond shallows, its radius continues to decrease. Lilies replace algae on its outer edges where water is too shallow for algae. As algae continues its growth towards the center of the pond, the growth of lilies follows. The pond continues to shallow. Algae continues its growth ahead of the lilies. Cattails grow on the outer edge of the lilies where the water is very shallow. Sedges follow cattails on the pond's edge where little of the pond water exists. Shrubs replace the outer edges of the sedges where the ground is now dry. Algae at the bottom of the pond in the center die as the lilies above them occupy the center of the pond. The lilies die and are replaced by cattails which have worked their way toward the center. Sedges replace cattails, and shrubs replace sedges over an area only a fraction the size of the previous pond. Finally, the climax returns with no sign of the previous pond remaining.

Plant successional stages occur on bare rock in a forest following glaciation or volcanism. Powdery colorful lichens, which are a combination of algae and fungi, grow on the rock and begin to dissolve it. These lichens are followed by a second sere consisting of colorful leafy lichens which continue to deteriorate the rock. Mosses of the third sere replace lichens, and the roots of the mosses continue to break the rock into fragments. Forbs and grasses invade the soil and germinate between the small particles of remaining rock fragments. These fragments continue to break up and form a fine soil. Shrubs replace forbs and grasses, each time increasing the amount of humus added to the developing soil. Finally the climax forest returns.

Plant succession accompanies fires in climax areas. Forbs are the first to germinate on the burned soils of a forest floor. These are followed by several seres, each with their representative species of grasses, shrubs, aspens, and climax forest trees, in that order. A large grove of subclimax aspens in a forested area is an indication of a previous fire or severe logging.

Forests have been used as an example where plant succession occurs, but these successions occur in tropical forests, grasslands, and other areas as well. Whatever the climax is, soils are improved, and seres gradually lead back to the climax vegetation. Such changes may require hundreds of years, depending on climate and other environmental conditions.

Fortunately, animals cope with these successional seres of plants. Certain species of animals are adapted to survive in certain seres. Horned lizards, starlings, house sparrows, jack rabbits, and deer prefer specific successional seres. On the other hand, grizzly bears, mountain lions, condors, and whooping cranes prefer climax areas. In other words, there is a sere for every species of animal, but in all cases, the existence of each sere is temporary.

BIBLIOGRAPHY

Raven, P. H., R. F. Evert, and S. E. Eichhorn. 1992. Biology of plants. 5th ed. New York: Worth Publishers.

Tony Gennaro 94

FUNCTIONS OF PELAGE COLORATION

Mammalian pelage coloration, or color of fur, plays one of two roles, depending on whether the animal is predator or prey. In either case, the name of the game is to hide. For example, a small mouse on the desert uses pelage coloration to hide from a predator. In contrast, a tiger in tropical areas uses pelage coloration to hide from prey, but only long enough to capture it.

Pelage coloration of predators or prey may blend in with environmental colors. Color in this case is for concealment, and the animal is said to be camouflaged. Examples include the tan fur with white spotting on fawns, stripes on tigers, spots on leopards, and the white coloration of polar bears. However, not all colors are for concealment.

Disruptive coloration is the name applied to pelage colors which break up the outline of an animal. The zebra is an excellent example. Black and white stripes on the zebra break the outline of its horse-like shape and consequently decrease the chances of the zebra being detected by predators. This effect can be quite beneficial to a zebra standing on the edge of the horizon in full view of an African lion. Stripes also break the outline of tails of raccoons and ringtail cats. These animals depend on their tails for balance as they climb trees and rocks in pursuit of prey. If the outline of this tail were not striped, its appearance would probably act as a warning device to prey. Also, stripes on the face of a badger may break the outline of the badger's head and enable the badger to approach prey close enough for capture.

Skunks benefit from their highly visible, contrasting black and white coloration. It is advantageous for skunks to advertise themselves, either to allow opponents to form an association between skunk odors and skunk colors or to be recognized by animals that have already made that association.

In some species, the function of pelage coloration is counter-shading. In this case, animals are darker on their top side than on their belly. Light from the sun or moon brightens the dark color on top of the animal, and at the same time, the absence of light on the shadowed underside of the animal causes that area to darken. The result is an even shade from top to bottom, and the animal blends with the surrounding environment. Such animals include rodents, squirrels, and rabbits. In contrast to counter-shading, certain species of rodents have the same shade of color on their top and bottom. These include species that are not exposed to sunlight and moonlight. Examples are voles and lemmings, which commonly remain in tunnels formed in the grass, and house mice and rats which remain in dwellings.

BIBLIOGRAPHY

Vaughan, T. A. 1986. Mammalogy. 3rd ed. Philadelphia: W. B. Saunders Company.

Tony Gennaro
94

THE GREAT WHITE SHARK

The New World Dictionary describes sensationalism as the "use of strongly emotional subject matter...intended to startle, shock, thrill, excite, etc." Frequently, species of wildlife are victims of media sensationalism, especially when they are considered a threat to humans. Authors of these emotional topics can be very effective in gaining the attention of an expansive audience, even though the author's comments may be untrue or grossly distorted.

Some victims of this kind of unfair sensationalism are—plants, fungi, molds, birds, bats, wolves, rats, snakes, spiders, and sharks. Sharks, especially the great white shark, get the most negative publicity. Great white sharks were already known to be dangerous to humans long before the film "Jaws" came out. Since then, their reputation went from big and dangerous to "mindless killing machines." Many marine biologists have indicated that "Jaws" was instrumental in encouraging humans to increase their kills of the great white shark. They say chances are good that this shark is declining in numbers, but they also admit that since they do not know how many existed in the first place, predictions of actual decline are impossible. A noted Australian big-game hunter stated that there are fewer great whites since the movie because so many have been killed by commercial fishermen who sell shark teeth. The going rate is about $3,000.00 for a complete set of upper and lower jaws with teeth. Furthermore, fishermen comment that catches of great white sharks in South Australia are down from ten to one or two a year, if any.

What is being done to reduce threats to the survival of the great white shark? Some individuals have elected to seek the truth about the way of life of these animals. The Cousteau research team is an example of such individuals. They include Jean-Michel Cousteau and Mose Richards who spent two years observing and tagging great white sharks, the largest carnivorous fish common to all seas. The research team tagged 67 sharks to study their movements, size increases, and weight gains. These research findings, though proven to be factual, are still struggling to overcome the damages done by sensationalistic writers.

Here are some discoveries made by the research team. Females give birth to 9 to 11 pups which are born alive. Each pup is approximately one meter long. Great white sharks are not loners as once thought. Several occur together, although they do maintain a fixed distance between one another, as would be expected of predators. Great white sharks establish a rank or order system among themselves, meaning the larger sharks have first choice of prey. Choice prey includes seals, squid, sharks, and carcasses of whales. Contrary to the beliefs of some individuals, humans are not on this menu.

The Cousteau team commented on the relationship between great whites and humans. As one member of the team put it, humans and great whites are neither friends nor enemies, but the general consensus is that the great white is not one that craves human flesh. These sharks will take an investigative bite occasionally, but consumption of humans is rare. For example, the coastline of California and Oregon reported 33 great white attacks in 53 years, and only four of those were fatal. In each case, the victim died of blood loss, not from being eaten by the great white shark.

Regardless of the new research done by Cousteau, most of the individuals who saw "Jaws" will choose to believe the sensationalistic film over the truth. They will retain the idea that great whites crave human flesh. This attitude will continue to be detrimental to the survival of the great white species. Like other top predators with few enemies, the great white shark had no need to produce large numbers of young per female in the past. Now, however, humans have become the shark's worst enemy, and newborns cannot be produced as fast as adult sharks are killed. The great white shark is in demand for sport, for its teeth and jaws, and for vigilante-like pursuits. Their very existence is now in jeopardy. At present, great white sharks are protected only off the coast of South Africa. Unfortunately, Australia and America have failed to act to protect this magnificent animal.

BIBLIOGRAPHY

Castro, J. I. 1983. The sharks of North American waters. College Station, TX: Texas A&M University Press.

Cousteau, J., and P. Cousteau. 1970. The shark: splendid savage of the sea. Garden City, NY: Doubleday and Company, Inc.

Dennis, F. Editor. 1975. Man-eating sharks: a terrifying compilation of shark-attacks, shark-facts, and shark-legends! Secaucus, NJ: Castle Books.

Ellis, R. 1976. The book of sharks. New York: Grosset and Dunlap Publishers.

Lineaweaver, T. H. III, and R. H. Backus. 1970. The natural history of sharks. Philadelphia: J. B. Lippincott Company.

Scharp, H., and M. Scharp. 1976. Shark wanted! dead or alive. South Brunswick, NJ: A. S. Barnes and Company, Inc.

Tony Brennan '94

INSECT

BEHAVIOR

Most of us have watched how insects behave. Some common observations include ants in a line carrying food from a source to their living quarters, bees pollinating flowers, and paper wasps building their nests. I have rested on my knees and elbows for hours on the desert sand watching the behavior of dung beetles. These insects work with dung from a grazing animal and form a ball about one-half inch in diameter. They roll this dung ball several yards, dig a hole, bury the ball, and then the female lays eggs in it. During my observations, I could not help but think that detailed planning and thought were involved in this dung-rolling process; however, science indicates otherwise.

Insect behavior is primarily innate or instinctive, meaning insects respond to a cue or stimulus. For example, the entire procedure among dung beetles from dung-ball formation to the laying of eggs is driven by responses to cues. The paper wasp is another example. A certain part of the wasp's nest may be unfinished. This unfinished part acts as a cue and brings about a response from the wasp to add more material to that area. Subsequent cues follow, and the wasp contin-ues to add material and respond to cues until the project is com-pleted. In other words, a wasp does not use a thought process to construct a nest in the same manner that we use in a thought process to build a house. Our ability to build stems from intellectual reasoning based on previous experience, from being told, or from reading in-structions.

We cannot say all insect behavior is entirely instinctive because some insects use both instinct and learning. Learning can be defined as a change in behavior brought about from experiences. Worker bees are good examples. They determine the direction and distance of nectar from the hive by responding to the body signals of a scout bee when it returns from a nectar source. However, because of their previous foraging activities, some worker bees are familiar with po-tential nectar sites in the area of the hive. Those bees find nectar sources earlier than bees that are not familiar with such sources.

Important as learning is, instinct cannot be considered an ineffec-tive form of behavior. Instinct is truly an asset for short-lived animals to conduct the numerous behaviors required to propagate themselves. Instinct is time-saving. Learning takes time, and it is most common among organisms with a long life span, such as humans and other mammals.

Finally, we ask, do insects display intelligence? Scientists say no. Intelligence involves reasoning. Reasoning exploits knowledge gained from experiences. The amount of knowledge required for reasoning is immense, and a large brain relative to body size is necessary for storage of that knowledge. A very elaborate nervous system is nec-essary to conduct the various kinds of behaviors typical of intellectual function. Insects lack these neural components which are supportive of intellectual capabilities. Therefore, scientists conclude that the behavior of insects is primarily instinctive, along with some learning, but they do not display intelligence.

BIBLIOGRAPHY

Alcock, J. 1993. Animal behavior. 5th ed. Sunderland, MA: Sinauer Associates, Inc.

Goodenough, J., B. McGuire, and R. A. Wallace. 1993. Perspec-tives on animal behavior. New York: John Wiley and Sons, Inc.

Starr, C., and R. Taggart. 1992. Biology, the unity and diversity of life. 6th ed. Belmont, CA: Wadsworth Publishing Company.

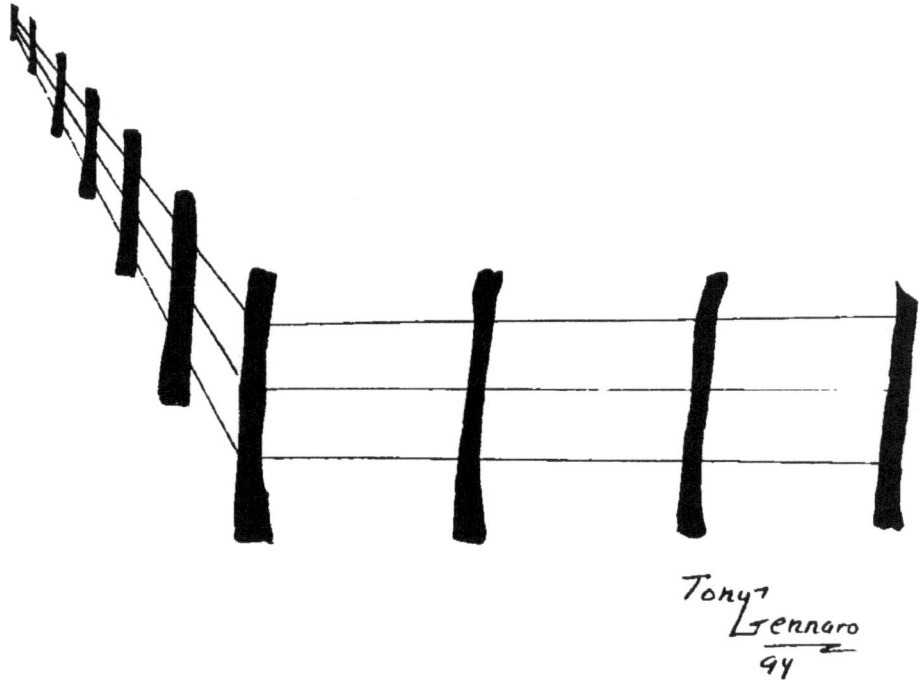

Tony Gennaro
94

TERRITORIALITY

Defense of resources is common in the animal world. Animals defend their homesites, food, water, kin, mate, and other materials of value against intruders of the same species. This process is called territoriality.

Direct attack of an opponent is one way to defend, but posting of boundaries is less dangerous and less energy-consuming. Humans commonly establish their territorial boundaries with fences. Many animals familiar to us have other means. Wolves and coyotes deposit urine on obstacles by raised-leg urinations. Cats spray urine on objects and drop fecal material. Grazing animals display threat postures. Birds advertise their boundaries by colorations and songs. If these boundary markers are not effective, then physical contact may be necessary. Usually this happens between males, although females do defend in rare cases. In most cases, the defender of a territory and the intruder do not fight to the death, as long as the loser of the bout has an avenue of escape.

This struggle for possession is not without benefits. Territoriality allows peaceful existence during mating and rearing of offspring. Breeding adults are free from invaders who may be seeking space, food, water, shelter, or a mate. There are other benefits for territorial species. For example, territoriality prevents overpopulation. Using birds as an example, let us assume that a particular area has only enough habitat for ten breeding pairs. If ten pairs capture space in that habitat and defend it, other pairs are excluded and usually will not breed. If these extra pairs do breed, it will likely take place in a habitat where chances are slim for survival of the offspring. Another benefit to breeding pairs in suitable habitat is that they usually have a large amount of space around their breeding sites. This dispersed pattern reduces the transfer of parasites and disease.

The most important benefit of territoriality is the defense of a mate. When a male defends a female, this behavior allows some insurance that his mate's offspring carry his genes. Therefore, the male's investment of time with a mate is of benefit to continue his heritage. More clearly, the energy he expends in defense of his mate and their resources goes toward the survival of a new generation consisting of both his and her genes. There is no way the male can ensure his transfer of genes into the next generation unless he defends the female continuously. The female benefits because she knows that her offspring possess her genes and that her heritage is secure. The male's only assurance for transfer of his genes is continuous observation and defense.

BIBLIOGRAPHY

Faaborg, J. 1988. Ornithology: an ecological approach. Englewood Cliffs, NJ: Prentice Hall.

Gill, F. B. 1995. Ornithology. 2nd ed. New York: W. H. Freeman and Company.

Vaughan, T. A. 1986. Mammalogy. 3rd ed. Philadelphia: W. B. Saunders Company.

Welty, J. C. 1988. The life of birds. 4th ed. Philadelphia: W. B. Saunders Company.

Tony Gennaro
94

BAT
ECHOLOCATION

Many kinds of bats echolocate, meaning they interpret features about an object in the dark by analyzing ultrasonic sound waves which travel from their mouth or nose to the object and then back to their ears. Bats accomplish this in the same manner as humans do when we hear our reflected voice in a canyon. We say hello and hear the words hello-hello-hello reflected off various structures. However, bats are much more thorough in their interpretation of reflected sounds. Bats interpret shape, size, and surface structures of objects, as well as whether the object is moving, its direction of movement, and the object's distance from the bat. The object could be a tree, rock, wall of a cave, other bats, or the bat's favorite food, the mosquito or moth.

Bats are especially talented at detecting and capturing moths in the dark. While searching, bats emit a vocal sound and wait for a reflection of that sound off a moth. The period of time between emission and echo is only a few thousandths of a second, and fortunately, bats are able to distinguish their echoes from the echoes of other bats. The interval of time between the emission and echo is very important. The brain of a bat evaluates the length of this interval to determine the distance of the moth from the bat. This interval gradually decreases as prey and bat get closer to each other. Furthermore, as the interval decreases, the pulse rate of emissions increases, and pulse-duration decreases. Finally, if the bat is lucky—and skillful—it makes contact with the moth and enjoys a delicious meal.

With this kind of equipment, it would appear that few moths can dodge a bat's pursuit. Actually, more moths escape than we would imagine, because moths employ countermeasures to avoid getting caught. They have sensory devices, like our ears, that are capable of detecting bat sounds. When moths hear these sounds, they make subtle, controlled escape maneuvers. However, if a high-speed bat happens to approach the moth within a detection range of about eight feet, the moth executes evasive responses, including power dives and loops to avoid capture. These erratic maneuvers are the final survival responses by the moth. If they are not effective, the game is over—the bat wins. Since the contest between bats and moths has been going on for thousands of years, we would expect each to improve their proficiencies, so that the most skillful bats catch, and the most evasive moths escape.

BIBLIOGRAPHY

Alcock, J. 1993. Animal behavior. 5th ed. Sunderland, MA: Sinauer Associates, Inc.

HORNS AND ANTLERS

A nimals endowed with horns and antlers also possess hooves and occupy grassland, shrubland, savanna, and forest-edge habitats. These animals are called ungulates. Horned ungulates include American pronghorns, African antelopes, rhinoceroses, sheep, goats, bison, and cattle. Antlered ungulates include deer, elk, moose, and caribou or reindeer. The latter two were once considered separate species, but recently, authorities have placed caribou and reindeer into one species.

Ungulates typically display herding behavior because many eyes, ears, and noses provide earlier detection of dangers. Once alerted, hoofed feet give ungulates the speed to outrun predators. When running is not effective, ungulates rely on their horns and antlers for a final defense.

These head ornaments may be used against any kind of threat, including their use as a challenge to members of the same species, particularly peers for rank during the breeding season. After all, the ungulate with the highest rank generally has the largest harem. In many ways, horns and antlers have similar functions, but these head structures differ in other ways.

One difference in horns and antlers is structure. Horns are composed primarily of hardened, thickened outer skin (epidermis) which contains keratin. Keratin is a protein which is the main component of fingernails, claws, talons, hooves, hairs, feathers, and reptilian scales. Only part of the horn, the very inner bony core, is not epidermis. This slender inner core is an extension of the ungulate's bony skull. Only the rhinoceros lacks a bony core. Antlers, on the other hand, are all bone, except for a thin outer layer of the epidermis, called velvet, which covers the antler for a short time only. Velvet quickly dies after the antler is formed, and this velvet flakes off or is scraped off by the animal itself, leaving behind the bony antler. Antlers on ungulates during the breeding season are composed of pure bone.

Horns and antlers differ with respect to sex of the ungulate. For example, both sexes display horns; however, only male ungulates bear antlers. One exception is the reindeer or caribou, in which case both sexes have antlers.

Horns and antlers differ in their growth patterns. For example, horns are never shed. American pronghorns, the only exception, shed the keratinized portion of their horns annually. On the other hand, all antler-bearing animals shed their antlers annually.

Regardless of differences in structure and growth patterns, horns and antlers are definitely useful tools for ungulate survival. Unfortunately, these head ornaments are useful to humans. These structures are used to make knife handles, belt buckles, and other ornaments. Horns and antlers are used also for medicinal purposes or aphrodisiacs. An aphrodisiac is defined as a substance which increases sexual desires. There is no evidence which supports horns or antlers as an effective cure-all, nor has it been proved that these structures have qualities as aphrodisiacs. Use of horns and antlers in the forms of ingestive products (preferably powders in pills or otherwise) as medicines or aphrodisiacs is considered to be traditional rather than functional.

Of special concern is that the use of ungulate horns for human needs is leading to the decrease in numbers of such species as the rhino. Rhino horns have an extremely high price on the black market (in some estimates, $12,000 per horn) for their use as carved dagger handles in North Yemen, aphrodisiacs in India, and cure-alls in

Southeast Asia and China. These demands have decreased rhino numbers from 100,000 to 11,000. Attempted measures to control rhino populations include elimination of poaching, reduction of the value of horns, restriction on imports and exports, and removal of horns. However, horns on rhinos grow back in one or two years. If these measures are not effective, there is no question among most authorities that the value of the horn will cause the demise of the wild rhino in about 13 years.

BIBLIOGRAPHY

Berger, J. January 14, 1993. Rhino conservation tactics. Nature 361:121.

Gunther, M. July/August 1993. Dehorning rhinos. Buzzworm Vol. 34.

Jackson, P. 1987. The rhino's fatal flaw. International Wildlife 17:4.

Modell, W. 1980. Horns and antlers, vertebrate adaptations: readings from the Scientific American. San Francisco: W. H. Freeman and Company.

Renate, S. 1989. Last chance for the black rhino. Alternatives 19:4.

BREEDING OF
SPADEFOOT TOADS

An old legend which folks in the arid Southwest commonly tell is that frogs drop from the sky during summer torrential downpours. This is said because frog sounds are heard everywhere following such rains. True, amphibian voices are heard following downpours, but these voices are being emitted by certain kinds of toads called spadefoots, not frogs. These spadefoot toads do not fall from the sky. Instead, these amphibians emerge from underground burrowing sites which they occupy, probably in a state of sleep, during the coldness of winter and dryness of summer.

Spadefoot toads leave their underground burrows only after heavy summer rains. This emergence is initiated by the sounds of raindrops as they impact on the surface of the soil. Burrowed spadefoots have such sensitive audio (hearing) mechanisms that they can perceive the heavy, continuous drops of rain on the surface of the ground, even though some spadefoots may burrow as deep as three feet. Once spadefoots reach the surface of the ground, they seek the nearest pools of water formed from the rains.

These pools are their breeding sites. Once males arrive there, they begin breeding choruses with vocalizations distinct to species. Females of the same species are attracted to these calls. This sexual recognition to species on the part of the females is important because usually more than one species of spadefoot occupies a pool of water at the same time.

Once the male and female of the same species get together following the vocal stimulus of the male, mating begins. The male mounts the female from behind, a term called amplexus, and sprays his sperm on her eggs as they are being laid. The calling and mating process lasts for about 24 hours. Then, all spadefoot toads depart from the breeding site, but not as pairs.

There is no need for pair-bonding because the fertilized eggs are left behind in the pond. In other words, there is no parental care in spadefoot toad behavior. The key to success of a new batch of spadefoots is rapid development of the tadpoles. These tadpoles develop into miniature spadefoots in 12 to 13 days. This is understandable because ponds usually dry in that period of time.

Young and adult spadefoots generally behave the same after they depart from drying ponds. Both age groups feed above ground as long as the soil is damp, and they depart to shallow burrows between feedings. However, as soils dry in summer or as cold temperatures approach in winter, a spadefoot will use its spade (a single sickle or wedge-shaped structure, one on the underside of each hind foot) to excavate a deep burrow. The rear end of the spadefoot faces the burrow entrance, and the toad actually backs into the hole as it is being excavated. Once a certain depth in the soil is reached, the spadefoot stops digging and remains there until the sounds of rapid summer raindrops are heard again. If raindrops are not heard during a dry summer, spadefoots will not emerge. Spadefoot toads remain underground that summer, the following winter, and await the sound of raindrops to occur their second summer underground.

BIBLIOGRAPHY

Goin, C. J., and O. B. Goin. 1971. Introduction to herpetology. New York: W. H. Freeman and Company.

McClanahan, L. L., R. Ruibal, and V. H. Shoemaker. March 1994. Frogs and toads in deserts. Scientific American 82-87.

Stebbins, R. C. 1985. A field guide to western reptiles and amphibians. Boston: Houghton Mifflin.

Wright, A. H., and A. A. Wright. Handbook of frogs and toads of the U. S. and Canada. 3rd ed. Ithaca, NY: Comstock Publishing Company.

Zug, G. R. 1993. Herpetology. New York: Academic Press.

Tony Gennaro
95

TERRITORIAL MARKING
IN WOLVES

To avoid conflict, owners commonly mark their property and guard it against intruders. This procedure is called territorial marking. It safeguards against thievery and allows property owners uninterrupted life functions. Humans declare ownership by establishing visual structures, such as fences, walls of dwellings, lettered signs, or other conspicuous features. Wolves, on the other hand, declare ownership by means of their odoriferous urine.

Boundaries of the wolf pack, the pack being a family of wolves and consisting of an alpha male and alpha female (the parents) and several generations of their offspring, are marked by raised-leg urinations of males. The urine is sprayed slightly above ground level on rocks, stumps, shrubs, or any other conspicuous area. Only the alpha male and other high-ranking members of the pack have privileges of marking.

The most intense marking is along the boundary of the pack's territory, where wolves from neighboring packs encounter each other's scents. Because of the seriousness of ownership, any neighboring wolf which crosses a territorial boundary will be chased off or killed by owners of the territory. In fact, neighboring wolves do not cross neighboring boundaries even if they are in pursuit of prey.

Less intense marking is conducted within the pack's territory. These markings are a means of communication between members of the pack rather than a means to delineate boundaries. Urine odors may last for 23 days; therefore, on the basis of age of the urine scent, a wolf may determine when another wolf marked the area. And, since the urine of each wolf has its own distinct odor, the sniffer might possibly identify the urinator as kin. This form of communication allows pack members to know the whereabouts of other family members when they are not in visual contact.

Territorial marking behavior of wolves is also conducted by domestic dogs which are descendants of wolves. Our family pet, a Boston Terrier whose name was Puggy, was a persistent marker. Puggy marked haphazardly every tree trunk, shrub, flowering plant, grass blade, and weed he could find in our yard. Furthermore, I was impatient with his persistence to mark every vertical object he encountered while I walked him in the neighborhood. I usually demanded that he remain tight on the leash and walk in a direction of my choice, not his.

One day, I decided to experiment with Puggy. I allowed him full rein to mark trees at will in a nearby wooded area. Puggy marked 23 trees, and during that process no obvious strategy was evident. In fact, the last five trees received a hiked leg, but no mark.

The marking behavior of Puggy and other domestic canines encouraged me to compare the efficiency of canine marking to the quality of writing. In that comparison, wolves compose clear, concise, meaningful, grammatically correct sentences; whereas, domestic canines scribble unclear, wordy jargon. If this comparison is true, the poor techniques of marking among some domestic breeds may result from the fact that they are restricted to smaller areas than those occupied by wolves, and these domestic breeds are not associated with a family pack. Domestic canines have not had an opportunity to learn good quality marking styles from experienced wild peers. Overall, territorial marking in domesticated canines seems to be an example of a capability which develops genetically, but which lacks the fine tuning acquired from learning.

BIBLIOGRAPHY

Allen, D. L. 1979. Wolves of Minong: their vital role in a wild community. Boston: Houghton Mifflin.

Hall, R., and H. S. Sharp. 1978. Wolf and man: evolution in parallel. New York: Academic Press.

Turbak, G. 1987. Twilight hunters: wolves, coyotes, and foxes. Flagstaff, AZ: Northland Press.

GILA MONSTER

Gila monsters were first discovered along the Gila River in Arizona. This animal is the largest North American lizard, the record length being 22 and 1/2 inches long. The gila monster's distinctive features are a halloween coloration and beaded skin. Halloween coloration refers to the patterns of black and dark brown blotches over a background of orange. Beaded skin means that the scales are rounded, rather than flat like the scales of most other lizards.

Indian folklore explains the color and beaded appearance of the lizard. Legend states that when the Tohono O'Odham Indians held their first saguaro cactus wine festival centuries ago, all the animals were invited. To look their best, gila monsters gathered bright pebbles and made themselves a durable and beautiful coat—a coloration they still wear today. The gila monster resembles no other lizard, except for the Mexican beaded lizard, which dwells in Mexico south of the range of the gila monster.

Despite their conspicuous appearance, gila monsters are rarely seen by humans. The reason is that these lizards hibernate during cold periods of the year. When they are not hibernating, gila monsters spend about one percent of their lives above ground. Energy demands of the lizards are low; therefore, they require only small amounts of food. A great deal of that food includes baby rodents captured in underground burrows.

This minimal, above-ground exposure has increased the survival rate of the gila monster. This is especially true since humans often capture and sell these lizards to collectors for a high price. Also, humans kill lizards out of fear. There is some merit for the fear of gila monsters because they are venomous.

The venom of gila monsters is used to reduce struggles of prey during capture. This venom is secreted along grooves of the lower teeth from glands in the lower jaw. Venom enters the wound of the prey made by the lizard's teeth. The quantity of venom in gila monsters is enough to kill small warm-blooded prey, but not enough to kill humans. However, gila monster bites should be cared for.

First of all, since the lizard clamps down with its jaws and holds onto the victim firmly, the only method of release is to pry open the lizard's jaws with a tool. Another person should assist in that task. The next step is to stop bleeding from the bite area. Then the wound must be cleaned and that part of the body containing the wound should be immobilized. Final treatment should include a tetanus shot, removal of tooth fragments, and an intake of pain-killers, antibiotics, and intravenous saline. Though folklore speaks of quick human death from the gila monster bite, patients are usually released from the hospital in a few days.

But not all gila monsters are aggressive, at least not the one I handled frequently for ten years. I named that lizard, Darwin, and I was always careful to hold Darwin gently around the neck just behind the head in case he had an aggressive idea in mind. However, Darwin never made one hostile move in my direction with his powerful jaws. The gila monster may be venomous, but this animal is not the "monster" its name implies.

BIBLIOGRAPHY

Brown, D. E., and N. B. Cormony. 1991. Gila monster, facts and folklore of America's Aztec lizard. Silver City, NM: High Lonesome Books.

HIBERNATION

One fall afternoon a youngster was removing a fence post from the ground. The soil surrounding the post was beginning to cool as a result of the gradual approach of winter. While digging, he uncovered the tunnel system of a small animal. Upon closer examination, he noted a small fuzzy creature curled in a nest chamber. The animal appeared to be dead, but following an attempt to uncurl the animal, it curled up again and initiated periodic short, quick-breathing actions. Not wanting to disturb nature, the youngster quickly returned the critter to its chamber and replaced dirt to the tunnel system. This unearthed mammal was assumed to be in hibernation, and it is likely that this assumption was correct.

Hibernation is a seasonal adaptation for animals which live in cold climates. Many mammalian species hibernate, including monotremes, marsupials, insectivores, bats, primates, and carnivores. These mammals hibernate to escape stresses of food shortage and unpleasant, cold winter weather. During hibernation, the metabolic rate of the body decreases; therefore, the body reduces its requirements for energy. With energy requirements reduced, the sleeping animal is able to survive on its resources of stored fat.

Hibernation among these mammals is initiated by certain environmental changes, or stimuli. The kind of stimulus involved depends on the species. Stimuli include low air temperatures, lack of food, decreasing photoperiods, and other factors, some of which are unknown. Responses to any one of these stimuli are numerous and may include a decrease in heart rate, constriction of blood vessels, decrease in breathing rate and oxygen consumption, and a suppression of shivering. Following these responses, the body temperature of the animal decreases (a process called hypothermia or torpor), body metabolism lowers, and the animal enters into a condition similar to sleep.

Even in hibernation, the animal maintains a neural alarm-system which enables it to cope with environmental changes that may be detrimental to survival. Certain parts of the brain are sensitive to noise, light, odor, and severe cold. Hibernating animals subjected to these types of events will awaken rapidly.

On the other hand, some hibernating mammalian species are responsive to gradual changes in environmental temperatures. For example, the body temperature of a certain species of mouse will decrease as environmental temperatures decrease, but only to a certain temperature. At that point, body temperature of the mouse remains stable as outside temperatures continue to drop. When environmental temperatures drop to minus five degrees Celsius, the mouse stabilizes its body temperature at two degrees Celsius. The mouse's body temperature never drops to a minus five degrees Celsius. If such a drop were to occur, the mouse would likely freeze to death.

Since energy is required to enter, as well as to emerge from hibernation, it would seem economical for animals to remain in hibernation until favorable seasonal conditions return in the spring—unless, of course, alarm stimuli are encountered by the hibernating animal. Even without alarm stimuli, however, hibernating animals awaken periodically. For example hibernating bears give birth during hibernation or awaken for reasons that are not detrimental to survival. Even mammals which undergo deeper stages of torpor than bears will awaken periodically.

The process of periodic awakening is part of a four-cycle phase of hibernation which includes the following sequence: 1) a normal active phase, 2) a 12-24 hour entrance phase, where body temperature drops, 3) a steady period of deep torpor that lasts from hours to weeks, depending on the species involved, and 4) an arousal phase, where body temperature returns to normal. Typically, a hibernating mammal may cycle through these phases many times throughout the winter season. For example, observations of several hibernating American badgers in Wyoming indicated that one badger entered torpor 30 times during a period of hibernation. However, even though this arousal was apparent, the badgers under study remained underground for 70 consecutive days. During this time-frame, these animals saved a tremendous amount of energy that they would otherwise have expended if they were active above ground.

BIBLIOGRAPHY

McKenzie, A. June 1990. Seeking the mechanisms of hibernation. Bioscience 40(6):425-427.

Vaughan, T. A. 1986. Mammalogy. 3rd ed. Philadelphia: W. B. Saunders Company.

Willson, M. F. 1984. Vertebrate natural history. Philadelphia: CBS College Publishing.

Tony Bennaro
95

AUTUMNAL
LEAVES

While walking through a wooded area in late September with Gary Pfaffenberger, my colleague, I asked Gary to recall his favorite event in nature. He answered, "This," as he pointed to colorful leaves on trees bordering our path. Gary stated, "Nothing in nature can ever be more beautiful than leaf colors in the fall of the year." He had a good point. Autumnal leaf colors are spectacular.

Leaf colors result from color pigments. Colors include green, red, purple, yellow, and gold. Each of these pigments is involved in the conversion of the sun's light energy into sugar for plant growth. A common question is, why do green pigments dominate color of plants in the summer and other leaf colors dominate color of plants in the fall?

The answer has to do with the abundance of green pigment, chlorophyll. Chlorophyll is present in much larger quantities than other pigments in early spring and summer. This intense quantity of green hides the presence of other pigments. Therefore, only the green pigment of plants is seen during the long days of the spring and summer months.

Long days and green colors are temporary in seasonal areas, that is, in the temperate zone where winter and summer alternate. As day length decreases in the fall, chemical changes occur within the plant. Chlorophyll breaks down, and the red, purple, yellow, and gold pigments reveal themselves in the leaves. The result is obvious. Within a matter of weeks, an explosion of beautiful autumnal colors appear. Unfortunately, this period of natural beauty is short because leaves soon fall.

There are reasons for leaf fall. Each leaf has a petiole which attaches the leaf to the plant's stem. A specialized layer of cells is situated between the base of this petiole and the stem. Chemical changes at the end of the growing season weaken this specialized layer, and the leaf breaks free.

While on the ground, leaves decompose and add humus to the soil. Humus functions as a fertilizer, retains moisture in the soil, and protects young sprouting plants. In turn, these plants grow, mature, and color the landscape with beautiful autumnal colors.

BIBLIOGRAPHY

Raven, P. H., R. F. Evert, and S. E. Eichhorn. 1992. Biology of plants. 5th ed. New York: Worth Publishers.

Stern, K. R. 1988. Introductory plant biology. 4th ed. Dubuque, IA: Wm. C. Brown Publishers.

Tony Lennaro
95

WOLF PACK STRUCTURE

The family is the structural unit of wolf society. A family of wolves is the pack which usually consists of 5 to 8 individuals, but packs of 36 wolves have been reported. The pack is composed of an adult breeding pair and their offspring. The breeding pair generally remains together for life.

The wolf pack is highly organized in that members are socially ranked. Rank among males is independent of rank among females. The leader of males is the alpha male. The leader of females is the alpha female. These alphas, or dominants, have control over activities of subdominants of their own sex. However, the alpha male is the leader of the pack; thus, he guides movements, initiates activities, and takes control at critical times such as during a hunt. This alpha male can be easily recognized because he is the only wolf among the pack that carries his tail high or straight out from the body.

Social ranking among pack members brings about peace and order, both of which are vital to the survival of the wolf family. For example, wolves of higher rank have first choice of food and other resources over wolves of lesser rank. This process of giving up resources reduces struggles among pack members.

There are inherent rules for breeding in the pack. The alpha male and female breed only with each other. Even though subdominant females come into heat, their breeding is a no-no. In fact, if a subdominant female receives any kind of sexual attention from subdominant males, she is attacked and severely punished by the alpha female.

Pups, on the other hand, are not ranked socially during their first year of life. They typically take liberties with elders by pushing, shoving, chewing, harassing, and climbing over them. The long-suffering elders often move away from the den to get sleep. At an early age, the young, woolly offspring get away with almost any kind of social behavior until they reach the end of their first year of life.

Restrictions on pup behavior by elders begin during the second year. At that time, elders become more demanding, snappish, and are ready to punish pups for immature behavior. Pups, which reach breeding age during their second year (22 months), remain with the pack, or if resources are abundant, they may disperse from the pack. For lone wolves, conditions are harsh, especially in the absence of the benefits of kinship. But, eventually one lone wolf finds another, and they pair with each other to form a pack of their own.

BIBLIOGRAPHY

Allen, D. L. 1979. Wolves of Minong: their vital role in a wild community. Boston: Houghton Mifflin.

Hall, R., and H. S. Sharp. 1978. Wolf and man: evolution in parallel. New York: Academic Press.

Turbak, G. 1987. Twilight hunters: wolves, coyotes, and foxes. Flagstaff, AZ: Northland Press.

THE AFRICAN LION CUB

African lion cubs are not easy to come by. Generally, a lioness will mate about 300 times with one or two males during each of her estrous (heat) periods. Each period lasts two to five days and on the average, only one estrous period in four leads to the conception of a litter of cubs. Therefore, a lioness must copulate at least 1,200 times to conceive, and at the most, only about one in four cubs ever reach maturity. This small survival rate would be less without the efforts of mothers of these cubs.

A pregnant female increases the odds of cub survival by giving birth in a secluded place, free from predators. This site is usually a few miles from her pride. The pride consists of her sisters, aunts, cousins, cubs, and a few males. The males are not kin to the pregnant mother, and one of them is the father of her cubs. The newborn cubs meet this pride at the age of six weeks when the mother leads the cubs out of their secluded birthplace.

These new pride members are innocent to dangers and are especially unaware of their slim chances for survival. Cubs begin their new life seemingly free of fear as they play with other cubs and adults in the pride. This play is important because it builds social bonds with other pride members, and it instills behavior the cubs will use as adults to kill prey.

Cubs are dependent on milk the first three months of their lives. They obtain milk from their mother as well as other lactating lionesses. After three months weaning begins, and cubs begin to eat meat which the mother brings from kills. This occurs only if prey is large enough for some meat to remain after the lioness satisfies her hunger. If her hunger is not satisfied, the cubs go without food. At five months or older, cubs follow their mother everywhere, and eat whenever food is available.

Pride members generally get along well at kills when prey is large and plentiful, but when prey is small and scarce, wrinkled snouts and barred teeth are common among those with prey in their possession. Only the largest and strongest eat. The result is that the smallest lions, the cubs, feed last. If food is scarce, cubs do not eat, and they slowly starve.

In his book, George B. Schaller describes the situation on the basis of his observation of lions in Africa. "Each summer as the dry season progressed, the pelvic bones of cubs grew more prominent. Their leg muscles shrank, making the cubs' paws look even more disproportionately large and ungainly. All play ceased. They plodded behind their mother, eyes vacant and hollow, with the hope of snatching a stray morsel should there be a kill. Toward the end of the dry season, the weakest cubs may stagger, and, lacking the strength to continue, they may lie down and watch the family retreat around a bend of the river without a backward glance. The cubs' lives then hang in the balance. If luck is with them a shower will bring forth green grass, which in turn attracts zebra. A few meals rapidly transform them into plump and vigorous youngsters, their period of deprivation forgotten. But if the drought continues there is no reprieve for some. Finally, unable to stand, their young lives retreat into the shadows."

Those cubs fortunate enough to reach an age of fifteen to eighteen months follow their mothers to kills less frequently. At two years of age, they reach subadulthood and have a good chance of survival. They are not good hunters at this age, and much of their activities are spent in trial and error. They rush at prey too early, fail to judge the speed of prey, and are unable to subdue a large animal. Schaller observed a young cub in its learning phase. He stated that the cub "snagged a gazelle during a communal hunt, perhaps her first kill. She dispatched it with a quick bite, then sat by the carcass with a surprised look on her face." That cub was successful; however, those less fortunate are given food by adults which serve as their back-up until they become more successful in hunting techniques.

At the age of two and one-half to three years, the grown-up youngsters become efficient at obtaining prey. Males leave the pride of their birth and seek a pride of their own. Females, on the other hand, usually remain and follow the cub-bearing and rearing behavior of their mothers.

BIBLIOGRAPHY

Bertram, B. 1978. Pride of lions. New York: Charles Scribner's Sons.

Guggisberg, C. A. W. 1975. Wild cats of the world. New York: Taplinger Publishing Company.

Nowak, R. M. 1991. Walker's mammals of the world. 5th ed. Vol. 2. Baltimore: The Johns Hopkins University Press.

Pusey, A., and C. Packer. August 1983. Once and future kings: groups of male lions compete for the chance to rule a pride and mate with its females. Their tenure is never secure. Natural History 92:55-57,60-62.

Schaller, G. B. 1973. Golden shadows, flying hooves. New York: Alfred A. Knopf, Inc.

CICADAS

Most of us recall those hot, dry, calm summer afternoons when bushes and trees came alive with loud, pulsating, buzzing sounds emitted in unison for long periods of time. An approach to the source of the sound to detect the noisemakers was of no value because—you guessed it—the buzzing stopped.

The buzzing sounds are emitted by adult male cicadas which are calling adult females of the same species. The intent of these males is to copulate with those female cicadas which respond to their calls. The sounds are created from vibrations of membranes situated on each side of the male's abdomen.

These highly vocal cicadas are also known by another name—locusts. Apparently, colonists at Plymouth, Massachusetts gave the name, locusts, to cicadas because colonists were reminded of the tales of large numbers of swarming locusts in the Bible. However, locusts are grasshoppers. Cicadas are very distinctive in appearance, and they do not even resemble grasshoppers. Adult cicadas, the only stage of the life-cycle we see, are large, winged insects with widely spaced, huge, protruding eyes. These insects reach one to two and one-half inches in length, and they resemble large flies.

The large size and relatively short life-span of adult cicadas (four to six weeks), in addition to their quiet nature when approached by intruders, explain why these conspicuous insects are more often observed dead than alive. When an adult cicada is observed dead on the ground, usually because it has reached the end of its longevity, the common statement of the onlooker is, "What on earth is that?"

In spite of this very short adulthood, the cicada has lived long enough to ensure the continuation of its species. This life-cycle is similar among other species of cicadas, differing only in length of time spent underground, anywhere from 2 to 17 years. The cycle begins as an adult male copulates with an adult female. Following that union, a female lays eggs in the bark of a small branch after she opens the bark with a saw-like organ, called an ovipositor, situated near the tip of her abdomen. Eggs hatch in two weeks. Newly hatched cicadas (called nymphs) drop to the ground and burrow down to the roots of the tree. Nymphs pierce the roots with sharp mouthparts and consume fluids from the tree's vessels. If the cycle lasts for 17 years, the nymph remains underground for that length of time. Upon emergence, grown nymphs leave open holes about three-fourths of an inch in diameter. Nymphs climb a tree, shrub, or any nearby object, shed their skin, and emerge as adults. In a short period of time, the wings of these cicadas unfold. Wings dry quickly and the cicadas become airborne. Open holes around the base of a tree or shrub and shed nymph skins on plants are evidence of cicada emergence.

Adult cicadas emerge annually. Each year 17-year cicadas emerge, and each one of those has served its 17-year term underground. Likewise, cicadas of other species with life cycles of various lengths emerge annually, and they, too, have spent the term of their nymph stage underground.

A question often asked is, "Do cicadas harm trees or shrubs?" Cicadas remove juices from plants, and branches die distally or outward from areas where eggs are laid, but these damages are not detrimental to the lives of plants unless large numbers of cicadas are involved. There is one case on record in the Midwest where severe tree damage was reported when 1.5 million cicadas emerged from one acre of ground.

BIBLIOGRAPHY

Arnett, R. H., Jr. 1985. American insects: a handbook of the insects of America north of Mexico. New York: Van Nostrand Reinhold Company.

Blaney, W. M. 1976. How insects live. London: Elsevier-Phaidon.

Brerenbaum, M. R. 1989. Ninety-nine gnats, nits, and nibblers. Urbana, IL: University of Illinois Press.

Clausen, L. W. 1954. Insect fact and folklore. New York: Macmillan Publishing Company, Inc.

Gillott, C. 1980. Entomology. New York: Plenum Press.

Ingham, K. July-August 1992. Cicada serenade. New Mexico Wildlife 37(4):10-11.

Matthews, R., and J. Matthews. 1978. Insect behavior. New York: John Wiley and Sons, Inc.

Swain, R. B. 1965. The insect guide. Garden City, NY: Doubleday and Company, Inc.

95 Tony Gennaro

BIOLOGICAL CLOCKS

Plants and animals conduct activities at specific times during a single rotation of the earth around the sun. This rotation is an event which takes 24 hours by our method of timing. For example, certain plant species open their flowers at specific times during the day; whereas, other species open their flowers at certain times during the night. Ground squirrels are active during the day, flying squirrels are active during the night, and male crickets chirp at the same time each evening. It would seem that these organisms are simply responding to light or darkness or some other environmental event; however, scientists suggest otherwise.

Scientists state that with each rotation of the earth around the sun, the environment displays predictable events. These include the onset of darkness in the evenings, the onset of light in the mornings, and other happenings. These occurrences are cyclic, that is, they repeat themselves each 24 hours—an environmental clock.

In the same way, scientists also conclude that plants and animals have an internal mechanism within the systems of their bodies called the biological clock. By chemical means, this clock initiates behavior and physiological activities at certain times during a 24-hour period, and these activities repeat themselves each 24 hours—like clockwork, so to speak. These events which take place at predictable times each day or each night are called rhythms. For example, the biological clock regulates the rhythms of flower opening, times of hamster activities, matings of certain species of fruitflies, and the times that crickets chirp.

It is obvious that rhythms of the biological clock must be synchronized with rhythms of the environmental clock; otherwise, day pollinated flowers would open at night. Such an occurrence would not be favorable because insects that pollinate these plants would be inactive at night.

Two questions arise about the environmental and biological clocks. Are the two clocks synchronized or are they independent of one another? If they are independent of one another, how are they synchronized? Research has been conducted to answer these questions.

Crickets were animals chosen for the tests. The males of this species typically begin to chirp at the onset of darkness. The first test dealt with clock independence, that is, does darkness (event on the environmental clock) cause crickets to chirp (event on the biological clock)? Crickets were removed from their ordinary cycle of 12-hours light and 12-hours darkness and placed into a chamber with 24 hours of continuous light. Chirping continued rhythmically (at the same time) throughout successive 24-hour periods of continuous light. This result proved the independence of the two clocks, meaning that darkness does not initiate chirping in crickets. However, while these continuous light experiments were being conducted, the onset of chirping began later and later during each 24-hour period. In other words, the time of the onset of chirping began to drift, indicating that the environmental and biological clocks were gradually losing their synchrony. However, when the crickets were removed from continuous light and exposed to a 12-hour light and 12-hour dark cycle again, the onset of the chirping shifted back to its original pattern and commenced soon after darkness. Thus, the two clocks, one environmental and the other biological, were again synchronized. Chirping, which was regulated by the biological clock, matched the onset of darkness which was regulated by the environmental clock. This is important because the female cricket makes an appearance at the onset of darkness. Obviously, chirps of the male cricket are conducted to attract the female for mating purposes.

Apparently the synchronizing cue for the two clocks is a change from light to dark on the environmental clock. This environmental cue is perceived by the crickets' eyes. The resulting impulse travels to the brain where the event, change from light to dark, and the event, chirping, are synchronized, and the result is that the two clocks run in harmony, meaning that the male cricket chirps at the onset of darkness.

There is good reason for the biological clock. Environmental events are fairly predictable. The biological clock prepares the body to anticipate and respond to these predictable events. In other words, the body's chemistry is cocked and ready for action each 24 hours.

BIBLIOGRAPHY

Alcock, J. 1993. Animal behavior. 5th ed. Sunderland, MA: Sinauer Associates, Inc.

Curtis, H. 1979. Biology. 3rd ed. New York: Worth Publishers.

© Tony Brennon 92

THE
HOUSEFLY

The fly, numbering about 100,000 species, is one of the smallest flying insects. One species, the housefly, *Musca domestica*, has been with humans as long as they have existed. This fly is recognized by four dark stripes, each about one-eighth inch long, running lengthwise on top of the segment behind the head.

Most individuals are not fond of houseflies or other species of flies with habitats similar to the housefly. Flies are not welcome in our home or on our food, and there is good reason for this attitude. The fly consumes and lays eggs in human and livestock fecal material and other waste products. It would not be so bad if they would stay put, but, unfortunately, they don't. As a result, we can expect that disease-producing organisms from the excrement will be transferred to foods.

Because of the close association between flies and humans, it is easy to understand why the housefly has been accused of transmitting a large number of diseases, including typhoid fever, dysentery, cholera, conjunctivitis, and others. No one really knows how guilty houseflies are, and only few studies support this conviction. But logically, the housefly's close associations with filth and attraction for humans make it a likely suspect.

In spite of this, some things can be said in favor of the fly. Probably a housefly's body is not a playground for disease-producing organisms because flies clean their bodies excessively. Plus, disease-producing organisms are constantly exposed to atmospheric drying and ultraviolet radiation. Now, if I ask you, "Do you still want a fly on your sandwich?" I suppose your answer would still be, "No." I agree, neither would I.

If we can get past the bad reputation and problems associated with the housefly, some of their unique features emerge. The life-cycle of a housefly from egg to adult is 10-20 days, with eggs being laid several times per year. Climate is an important factor which determines how many times eggs are laid. For example, the number of generations of eggs increases with warmer climates. If all the offspring from a single pair of houseflies survived, in one summer these offspring could produce a layer of flies 47 feet high over an area about the size of Arizona and New Mexico combined. Now, that is prolific.

Another unique feature is the fly's mouthparts. These are on the end of a tubular organ which remains withdrawn in the fly's head when the fly is not feeding. When the fly locates a likely meal, this organ telescopes downward, in the same way we would extend the legs of a tripod, and fleshy lobes on the end of this organ puff out. Imagine this being similar to an air bag being inflated within a vehicle. At the same time the mouthparts are being extended, the fly vomits on the food to facilitate digestion. These lobes then press down on the vomit and food which are sucked up into openings on the underside of the lobes. Isn't that a delicious thought? These puffy fly lips are designed to filter materials which enter the mouth, so even on excrement, flies have a choice of acceptance or rejection.

Unique features aside, because of their habits and general peskiness, the numbers of flies must be controlled. Prevention seems to be the way. For example, proper sewage disposal, sealed garbage containers, screened windows and doors, disposal of pet excrement, and effective garbage collection are important. Prevention is the most effective way because the housefly easily becomes resistant to insecticides. When all is said and done, there is still one very effective way to kill flies—the S and S method, that is STALK and SWAT—with a fly swatter.

BIBLIOGRAPHY

Conriff, R. July 1989. Why God created flies. Audubon 91(4):82-85.

Oldroyd, H. 1964. The natural history of flies. New York: W. W. Norton and Company.

BROOD PARASITISM

Few people have heard of brood parasitism. It is a process whereby certain species of birds, called brood parasites, lay their eggs in the nests of other species of birds. In turn, the owner of the parasitized nest, the host or foster parent, incubates and rears the brood of the parasite. Brood parasitism is not a common event in the bird world.

Only one percent of all species of birds are brood parasites, and almost all species of brood parasites within that one percent belong to a group called cuckoos. Only 50 of the 150 species of cuckoos are parasitic.

One well-known brood parasite is the European cuckoo. This species mates frequently, but the female's eggs do not mature until she sees the nest of a potential host. Once that happens and egg maturation begins, the female cuckoo cautiously approaches the nest to be parasitized, removes one egg from the host's nest, and lays one of her own among the host's eggs. The host does not recognize the newly laid egg because it is similar in appearance to her own eggs.

Other behavioral features are unique to brood parasites. The European cuckoo can lay an egg in five seconds. This fast-laying strategy decreases her chances of being seen in the host's nest. She lays only one egg for perhaps a couple of reasons. The young cuckoo which hatches from the egg will not have to compete with other cuckoos, and one extra egg in addition to those laid by the host may go unnoticed. The young cuckoo usually hatches first, and it forcibly removes the host's eggs from the nest. The cuckoo does this by placing its body under the egg and positioning the egg in a shallow depression on its back. The cuckoo then raises its body to the rim of the nest, heaving the egg over the side. If a host nestling hatches from an egg, it has a short life because the cuckoo will also heave it over the side of the nest. The result is that the young cuckoo ends up with the entire nest to itself and will be fed and reared by the host, even though the young cuckoo will quickly reach a larger size than the host.

The advantage of host parasitism is obvious: it enhances the reproductive success of European cuckoos. A female cuckoo lays one egg in each of 28 different host's nests each breeding season. The survival rate of cuckoos raised alone is expected to be higher than 28 cuckoos raised together in one nest cared for by real parents.

BIBLIOGRAPHY

Pettingill, O. S., Jr. 1985. Ornithology in laboratory and field. 5th ed. Orlando, FL: Academic Press, Inc.

Van Tyne, J., and A. J. Berger. 1976. Fundamentals of ornithology. New York: Wiley-Interscience Publication.

Wallace, G. J., and H. D. Mahan. 1975. An introduction to ornithology. 3rd ed. New York: Macmillan Publishing Company.

Welty, J. C. 1988. The life of birds. 4th ed. Philadelphia: W. B. Saunders Company.

Tony Gennaro
94

COCKROACHES

W hat is the most detested insect to be found in a home? It is flat-bodied, tan to brown in color, slippery, and quick to run for shelter when the lights are turned on? Answer—the cockroach. It is likely that all humans have encountered cockroaches at one time or another, and for that reason, humans have targeted them with 25% of insecticide use in the United States. Other questions about cockroaches come to mind. Are cockroaches native to the United States? How many species of cockroaches inhabit the United States? How many species of cockroaches occupy this planet?

About 3,500 species of cockroaches have occupied this planet for 300 million years, and most of them live primarily in rain forests. Only five species are common household pests, and all five were accidentally introduced into the United States.

All five species have similar life-cycles. Adults feed and mate. The female produces several rectangular-shaped egg cases in her lifetime. She deposits these cases in suitable areas, and soon immature cockroaches, called nymphs, emerge from them. Nymphs feed, grow, and shed their skins several times until they reach the adult stage. The life cycle begins again as adults mate and produce more cockroaches. The length of this cycle depends on temperature, relative humidity, and the species of cockroach.

All five species of house-dwelling cockroaches are quite comfortable with living conditions suitable to humans. These conditions include available water, temperatures above 65 degrees Fahrenheit, starchy and sweet foods, meat products, and edible substances preferred by humans or their pets. Cockroaches live comfortably with humans, but the opposite is true for humans. Why do humans dislike cockroaches so much?

Some reasons are that cockroaches are unattractive, appear unexpectedly, emit unpleasant odors, are suspected of transmitting diseases, and give the psychological effect of uncleanliness. However, these flat, unwelcome insects may not deserve all the bad press concerning spreading of disease.

There is evidence that human allergies may be caused by cockroach habits. Minute particles from shed skins, egg cases, and fecal droppings of cockroaches cause allergies among humans. However, no evidence shows cockroaches to be disease carriers. In fact, some researchers indicate that the pads of cockroach feet secrete anti-bacterial substances that prevent the transmission of disease-producing germs.

Cockroaches have uses to scientists who use them for behavioral and medical experiments. Cockroaches are subjects for testing the effectiveness of insecticides on household insect pests.

Despite their uses in scientific research, most humans agree that free-running cockroaches in human dwellings must be controlled and there are various means for doing this. Some methods include: control all water leaks, seal all cracks and openings which allow cockroaches to enter dwellings, and keep feeding areas of pets clean. Vacuum rugs in eating areas frequently, and most importantly, place all trash containers in open areas where cleanliness can be maintained. When these preventive measures fail, scientists have provided more effective means of control.

Various chemicals for cockroach control are widely available. Some are poisons developed to kill cockroaches on contact. Others, referred to as pheromones, emit odors similar to those secreted by one sex of cockroach and which attract members of the opposite sex. Pheromones are harmless, and they are used as lures in traps. Once cockroaches enter a trap, they step on sticky material and cannot escape. It is likely that any one of these methods is more effective than hand-to-hand combat, that is, smashing cockroaches one at a time during their temporary, startled, motionless state which follows when a light is turned on.

BIBLIOGRAPHY

Alford, A. R. September 1991. Could the cockroach prove to be our best friend? USA Today 19:80-82.

Arnett, R. H., Jr. 1985. American insects: a handbook of the insects of America north of Mexico. New York: Van Nostrand Reinhold Company.

Comstock, J. H., A. B. Comstock, and G. W. Herrick. 1931. Manual for the study of insects. Ithaca, NY: The Comstock Publishing Company.

Hickin, N. E. 1974. Household insect pests. New York: St. Martin's Press.

Metcalf, C. L., and W. P. Flint. 1951. Destructive and useful insects: their habits and control. 3rd ed. New York: McGraw-Hill Book Company, Inc.

Wilson, M. C., G. W. Bennett, and A. X. Provonsha. 1977. Insects of man's household and health. Prospect Heights, IL: Waveland Press, Inc.

THE FEMALE
AFRICAN LION

Female African lions live in an extended family unit called a pride. They share this unit with a coalition of two or three adult males, several litters of cubs, and other females which are their sisters, aunts, or cousins. The adult male coalition remains with the pride about three years until it is chased out by a coalition of younger males. Male cubs leave the pride at a subadult age of about three years. Therefore, the average tenure of males in a lion pride is short, but this is not true for the females.

Adult females remain with the pride, and their female cubs grow up to be lionesses which, in turn, spend their lives in the pride. Those females that leave as subadults do so because the pride may be attaining maximum size, or females may fail to bond closely with the other adult female residents. In any event, as female cubs mature, they replace aging lionesses and maintain a breeding and bearing status for about ten years. Finally, adult females spend their last few years in the pride as nonreproductive, noncontributory elders before they die, usually at an age of 15 years.

Obviously, adult breeding females are the basic foundation of pride existence. They encourage copulatory behavior from males. Females give birth to cubs, nurse them, and develop bonds between cubs and other pride members. Females assist other lionesses in food capture for all members of the pride, including food consumed by adult males.

Females benefit from males in two ways. Since males guard the edges of the pride's territory against intrusion from nomadic lions, female functions are conducted with minimal disturbance. More important, the female obtains an input of fresh genetic material from males of the coalition which were born elsewhere in another pride. This genetic material is combined during conception. Thus, cubs are born with the genetic variability essential to survive in a harsh environment. Female cubs which survive and mature repeat cycles of the previous generations of mothers.

BIBLIOGRAPHY

Bertram, B. 1978. Pride of lions. New York: Charles Scribner's Sons.

Guggisberg, C. A. W. 1975. Wild cats of the world. New York: Taplinger Publishing Company.

Nowak, R. M. 1991. Walker's mammals of the world. 5th ed. Vol. 2. Baltimore: The Johns Hopkins University Press.

Pusey, A., and C. Packer. August 1983. Once and future kings: groups of male lions compete for the chance to rule a pride and mate with its females. Their tenure is never secure. Natural History 92:55-57,60-62.

Schaller, G. B. 1973. Golden shadows, flying hooves. New York: Alfred A. Knopf, Inc.

TICKS

Almost everyone is familiar with ticks, and their unpopularity is understandable. Ticks are dangerous to the health of humans mostly because ticks feed on blood. With this kind of liquid diet, a vector, in this case a tick, establishes a blood connection between its tissues and the tissues of its victim, the host. The host of a tick can be any warm-blooded animal.

This liquid pathway between vector and host enables disease-producing organisms in the tick to enter tissues of the host. Once inside the body of the host, these organisms initiate various diseases which include Rocky Mountain spotted fever, relapsing fever, tularemia, Colorado tick fever, Q fever, and Lyme disease. It is their disease carrying ability that makes ticks a concern to humans.

Ticks are very specialized at obtaining their blood meal. They insert barbed mouthparts into the host's flesh and draw blood by means of a specialized feeding tube. Because this procedure is not painful to the host, the tick can feed undisturbed. Their small size and ability to pierce skin painlessly allows ticks to avoid detection.

The small size of ticks is demonstrated by the species, *Ixodes dammini,* which transmits Lyme disease to its hosts. The adult stage of this tick is large enough to cover the four numbers on the date of a dime. A stage in growth prior to the adult, the nymph, covers two numbers of the date. The larva which grows into a nymph covers the top half of the number 9.

These small insects are difficult to detect in their favorite habitat which typically includes wooded, bushy areas where they are most frequently encountered during April through October. People in tick-infested country are definitely alert for ticks during those months.

The life history of outdoor-dwelling ticks is similar for most species. Adults attach to mammal hosts and feed continuously until they engorge themselves with blood. Then they drop from the host, and the female lays her eggs. Larvae hatch from eggs and move onto blades of grass. Larvae remain there until they come in contact with passing hosts. Then, larvae cling to hosts and feed on their blood. Soon these larvae drop from the host and change into nymphs. Nymphs move from vegetation to host for their fill of blood. Nymphs leave the host to molt into adults and the cycle begins again. In some species, this cycle may take several years to complete.

Ticks have one main purpose in life—locate a host and feed on its blood. Therefore, a likely host, such as a human, should avoid contact by tucking in pant legs and shirttails, wearing long sleeves, and applying tick repellent. If contact is made and the tick is located, it is advisable to remove the tick as soon as possible.

The following procedures for removal are in use. They include applications of various substances to the tick which may encourage it to loosen its grip. Materials include nail polish, alcohol, petroleum jelly, chloroform, and heat from matches or cigarettes. However, the best means of removal involves grasping the tick as close to the skin surface as possible with forceps or tweezers. If fingers are used, they should be protected from the tick's disease organisms with rubber gloves or tissue paper. A gentle pull will generally remove the tick. Then, the tick should be disposed of in alcohol or washed down a drain. There is no concern if the tick's mouthparts remain on the skin, although a festering sore may occur at the insertion area for a short period of time. The main thing is to remove the body of the tick.

Potential hosts can never assume that they have evaded the lurking capabilities of a tick. Departure from tick-infested areas should always be accompanied by a thorough inspection of all parts of the body, including the bodies of pets. Odds are that those who trek through tick country will have a tick or two on their body which will need to be removed.

BIBLIOGRAPHY

Bennett, W. I. June 1985. Of ticks and men (and women and children). Harvard Medical School Health Letter 10:1-4.

Collins, P. J. August 15, 1985. Practical briefings. Patient Care 19:24-25.

Dogel, V. A. 1964. General parasitology. Edinburgh: Oliver and Boyd.

Dowie, E. D. July-August 1984. Beware the tick! Mother Earth News pp. 45.

Hoeppner, G., D.V.M. July-August 1990. What makes fleas flee and ticks tick? The Saturday Evening Post 262:18-20.

Marzouk, J. B., M.D. June 1985. Tick born diseases: where to expect and how to detect such bite-caused syndromes. Consultant 25:21-36.

Miller, G. September 1982. Body language: pint-sized pests. Backpacker 10:87-89.

HUMAN

VS.

CETACEAN DIVERS

Following a deep dive in water, humans can experience severe pains upon returning to the surface too quickly, and the resulting body injury may cause unconsciousness and death. Aches are so severe in some cases, commonly in the joints, that divers bend over in agony. Hence, the name, the bends. However, the more common name for this condition is decompression sickness, meaning that the sickness results from decreasing pressure.

Divers experience decreasing pressure upon ascent to the surface of the water. This is expected because pressure gradually increases with increasing depth of the water. Increases in pressure with depth are a result of increasing weight of the water. For example, at 150 feet, because of this weight, pressures on all surfaces of the body are four and one-half times greater than they are at the surface. At this depth, pressure on the diver's chest cavity, lungs, and air within the lungs causes gases such as nitrogen in that air to dissolve in the blood of lung tissue. Dissolved nitrogen at high pressure and deep water causes no problems as long as the diver does not ascend. When the diver ascends, pressure decreases (decompression), and nitrogen is released from the blood into body tissues as bubbles. This process is similar to the appearance of bubbles in a soda after the cap is removed. Bubbles appear because the pressure that kept them dissolved was removed with release of the cap. These nitrogen bubbles in tissues cause damage, sickness, and death. However, some mammals do not experience decompression sickness. These are the whales, porpoises, and dolphins, known collectively as cetaceans.

Cetaceans do not experience complications during dives as humans do, even though nitrogen dissolves in the blood of cetaceans to the depth of 200 feet. Beyond that level, cetacean lungs collapse and air within them is shifted to various body spaces where gases fail to dissolve in the blood. Why don't cetaceans display decompression sickness when they ascend from a 200-foot depth?

Authorities give this answer. Because cetacean heart functions are smooth and quiet, shock waves are minimal in cetacean body systems. Therefore, gases escape from the blood slowly without creating bubbles. Contrary to cetaceans, human heart functions are not smooth, and considerable shock waves are created which cause gas bubbles in the blood, and thus, decompression sickness.

Unlike cetaceans, humans are not endowed with means to cope with deep diving in the water. Humans have, however, developed some guidelines which, when followed, prevent decompression sickness. These guidelines include mathematical calculations which instruct divers on how long to remain submerged at certain depths and how to avoid absorbing excessive amounts of nitrogen. Another recent development is the underwater computer which divers wear attached to their gear to provide information on durations of stay at various depths and surfacing procedures.

BIBLIOGRAPHY

Burton, R., M.A. 1980. The life and death of whales. 2nd ed. New York: Universe Books.

Coffey, D. J. 1977. Dolphins, whales and porpoises: an encyclopedia of sea mammals. New York: Macmillan Publishing Company, Inc.

Gaskin, D. E. 1972. Whales, dolphins and seals: with special reference to the New Zealand region. Auckland, New Zealand: Heinemann Educational Books.

Martin, K. 1988. Giants of the sea. New York: Gallery Books.

Martin, R. M. 1977. Mammals of the oceans. New York: G. P. Putnam's Sons.

SOURCES OF WATER FOR WILDLIFE

Plants and animals cannot survive without water because it is essential for chemical reactions within the cells of all living things. Therefore, plants and animals must obtain water regardless of the areas they inhabit.

Considering animals, one wonders how wildlife species survive on deserts and grasslands where there is no sign of visible water. Such wildlife include insects, spiders, lizards, rodents, birds, antelopes, foxes, coyotes, and many others. How do they do it?

Some have special techniques. Take pebble collectors for example. These Australian mice build a water source outside their burrows by covering the entrances with even-sized small pebbles. Then, these mice obtain water by drinking dew which collects on the pebbles.

Other species use different methods to obtain water. Some animals drink dew from plants. Others consume water by eating succulents which are plants that contain large amounts of water in their cells. The pack rat is especially adapted to retrieving water from cacti which are very important succulents in deserts. Other animals, such as foxes, obtain water from prey which is composed of about 70% water. However, no animal in deserts or grasslands is as specialized at obtaining water from its environment as kangaroo rats.

Kangaroo rats use metabolic water. This water is produced chemically from hydrogen atoms which the rodent receives from carbohydrates and oxygen atoms which it receives from the atmosphere. Carbohydrates are the major components of seeds which the kangaroo rat consumes. Therefore, these rodents survive solely on a diet of dry desert seeds, and they have no need to consume free water, succulents, or prey animals.

Many species display behaviors which conserve water. For example, some animals seek shade under trees, bushes, and rocks during high temperatures of the day. The coolness of shade reduces evaporation of water from the body. Shade is used to the maximum by animals that are active during the night. At that time, the earth shades animals from the sun. Such animals are said to be nocturnal, and kangaroo rats are the best example. During the day, kangaroo rats remain in their burrows where it is cool and humid. During the night, they are active on the surface where it is cool and humid. Although kangaroo rats and several other species of rodents occupy the hottest deserts on this planet, because of their behavior, they never really expose themselves to true desert conditions.

BIBLIOGRAPHY

Tate, R. 1971. Desert animals. New York: Harper and Row.

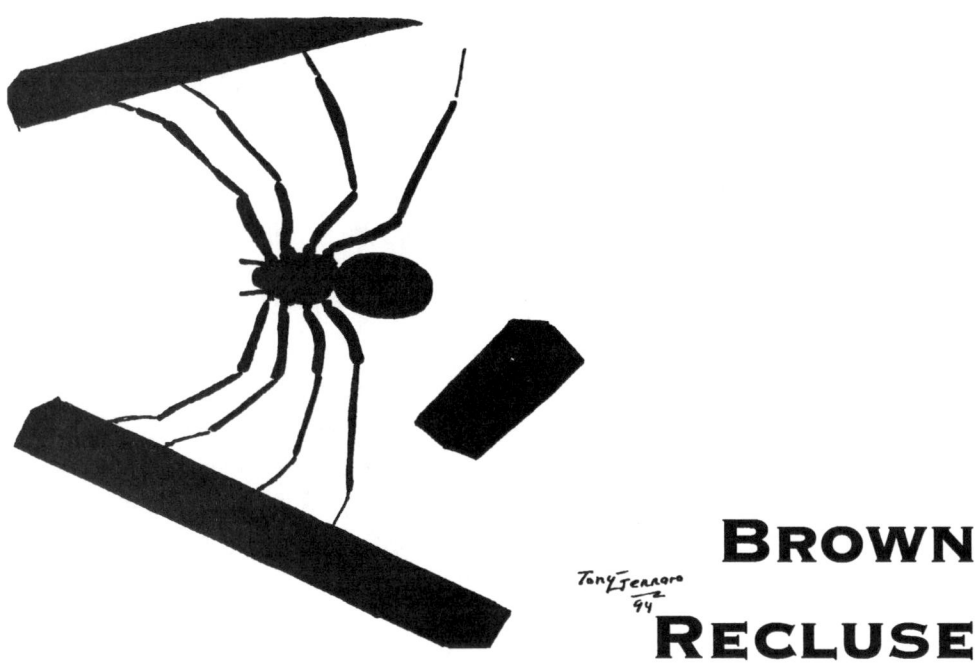

Tony Jennaro
'94

BROWN
RECLUSE

Be alert for a spider, about one-half inch in size, yellowish to dark brown in color, long legs, and a violin-shaped marking on top of the cephalothorax. The latter is the fused head and thorax of the spider. The abdomen behind the cephalothorax is equal in size to the cephalothorax and has no markings.

The features described are those of the brown recluse spider which is called other names, such as fiddleback, violin spider, or brown spider. Both sexes of this spider are very venomous and dangerous to humans. There are 100 species worldwide, and 13 of them are in the United States. The most common is the brown recluse, *Loxosceles reclusa,* which inhabits Hawaii and the United States from New Jersey west to California. It is most common in midwestern United States.

Throughout the region inhabited by brown recluse spiders, they prefer dark, remote areas—hence, the name recluse. In southern United States, brown recluse spiders occupy dwellings, but they prefer the outdoors under rocks, woodpiles, and debris. In northern areas, these spiders prefer indoors, where they occupy attics and storage areas. Unfortunately, all of the aforementioned areas are frequented by humans; therefore, the brown recluse is a serious threat, especially to children and feeble adults.

The bite of the brown recluse causes extensive tissue damage and severe reactions of the inner body. Tissue damage at the site of the bite can encompass an area one inch or more in diameter. Scarring may occur, and plastic surgery may be necessary, especially if the damage is on the face. Reactions to the bite include fever, chills, weakness, nausea, vomiting, convulsions, damage to blood cells, and kidney failure. These body malfunctions could lead to death. With such body damage, the obvious concern is treatment.

Unfortunately, there is much to be discovered about treatment for brown recluse bites. No medication prevents tissue damage at the site of the bite, few remedies relieve systemic reactions, and there is no evidence that the antivenom is effective. So what can one do?

Be alert. Be vigilant. Prevention is the remedy. Shake out foot-gear, clothing, and blankets before using them indoors and out, especially in midwestern United States. Try to capture the spider which caused the bite to help authorities identify it. Treatments for each species of spiders have important differences because spiders, other than brown recluses, cause tissue damage, but they do not cause systemic reactions and death. Therefore, recognition of the spider is necessary to assist personnel during treatment at a medical facility.

BIBLIOGRAPHY

Russell, F. E., M.D. and Ph.D. May 1991. Arachnid envenomation. Emergency Medical Services 20:16-22,46-47.

Tony
Grennaro
94

THE BARK SCORPION

Visualize a scorpion. It usually has a glassy appearance and is black, tan, or yellowish in color, depending on the species. The forward part of its flattish body contains eyes, four pairs of legs, and a pair of extended pincer-like appendages. Behind the forward part, the abdomen tapers gradually rearward to form a slender tail which ends in a sharp, pointed stinger. The stinger is used to inject venom into victims for prey capture and defense. No other animal fits the description of a scorpion, and no little creature is feared more by humans. Not all scorpions, however, are worthy of such fear.

Only one species, the bark scorpion, *Centruroides sculpturatus*, is lethal to humans in the United States. This scorpion, which occupies arid and semiarid habitats in the U.S., is common in Arizona and rare in Texas, New Mexico, California, and Nevada. The bark scorpion is straw-colored and one-half to three inches in length, depending on age. This scorpion is distinguished from other species by slender pincers, each about six times as long as they are broad and a rectangular, rather than square, last segment on the tail. The shape of the bark scorpion's stinger is similar to the shape of stingers on other scorpions.

The sting of the bark scorpion causes serious problems for adults, but children and the elderly are more severely affected. Symptoms range from pain at the site of the sting, blurred vision, excessive salivation, problems with swallowing, slurred speech, difficult breathing, to involuntary jerking of arms and legs. There is treatment for the sting of the bark scorpion, and victims have a good chance for survival.

Treatment involves the use of an antivenom which eliminates symptoms within one and one-half hours. However, patients must be observed continuously following treatment to monitor heart rate, body temperature, breathing, and seizure-like symptoms.

As a precaution, it is advisable for anyone stung by a scorpion to seek medical assistance. And, during that time, decisions about treatment are simpler if the scorpion in question is available for identification.

However, as is true with all excursions into nature, preventive measures are always the best means to avoid contact with venomous wildlife. For example, humans should be especially aware of habitat preferred by the bark scorpion, especially in Arizona where that species is common. The bark scorpion frequents trees, such as the sycamore, mesquite, and cottonwood, and it prefers being under bark of old stumps, under rocks, in lumber piles, or in firewood. Bark scorpions crawl into footgear, bedding, and clothes. Therefore, a word of advice to individuals in bark scorpion country—shake out those boots first thing each morning before slipping them on because bark scorpions are especially active at night.

BIBLIOGRAPHY

Curry, S. C., M.D., M. V. Vance, M.D., P. J. Ryan, M.D., D. B. Kunkel, M.D., and W. T. Northley, Ph.D. 1984. Envenomation by the scorpion *Centruroides sculpturatus*. Marcel Dekker, Inc.

Tony Ferraro
95

FATE OF AFRICAN
ELEPHANTS

African elephants are in trouble. Most authorities predict that these elephants will become extinct by 2010. The basis for this prediction is that elephants are being killed at a greater rate than they are reproducing. This loss of elephants may take place despite all efforts to save them. Their problems are two-fold: the high price for their tusks and competition between the African elephant and humans for space.

Ivory from African elephant tusks which is used to produce figurines, jewelry, and other items has had a high market value for many years. Because of the demand for products from ivory, the price of ivory reached a peak in the 1970's. At that time, costs rose from $5.00 to $50.00 per kilogram (2.78 pounds); consequently, the killing of African elephants increased. Conservation organizations responded to elephant kills by placing these large mammals in threatened and vulnerable categories.

Despite good intentions, efforts to stabilize elephant populations failed, and the market value of ivory continued to rise. Ivory reached $100.00 per kilogram by the 1980's. Eighty percent of the African elephant kills for ivory at that time was illegal, and, after Japan, the

United States was the largest importer of those illegal tusks. In fact, in 1986, ivory carvings shipped to the United States represented 32,000 dead elephants, and in that same year, the United States imported raw tusks and skins from 11,000 other elephants. Those transactions resulted in an annual retail value of $100 million for African elephant products in the United States.

Ivory remains in demand, and the tropical forest and savannah habitats of the elephant continue to be reduced by agricultural activities of an exploding human population in areas where elephants once roamed freely. Conflicts between humans and elephants have restricted seasonal movements of elephants. Those restrictions have led to habitat destruction by African elephants in areas where they have been confined.

National parks are the only areas remaining where elephants can function without friction with humans. These refuges, however, do not provide enough space, food, and water for expanding protected elephant populations. Therefore, these animals are being culled in those parks, even though such activities are objectionable to many individuals.

Destruction of African elephants everywhere has promoted attempts to stabilize their populations. In 1990, for example, most African elephant populations were tagged as endangered. The United States and other European communities banned the import of all ivory products in 1989. Efforts are being made to use profits from legal ivory trade to finance the conservation of elephants. However, those sales are disapproved of by individuals who predict that such endeavors will only maintain the high price for ivory.

Despite all efforts to save African elephants, there seem to be only two possible solutions for their survival: one is to reduce the price of ivory below favorable profit margins, and the other is to stop humans from encroaching upon tropical forests and savannahs. Neither solution seems likely at this time.

While no one wants the largest terrestrial mammal on planet Earth to be in their backyard because of the destruction that animal would cause in a single day, the activities of an elephant in its own backyard is a valuable asset to ecology. Seasonal movements of African elephants create open areas in dense woodland and jungle and reduce the spread of shrubs. Open areas allow the penetration of sunlight to otherwise dark forest floors. Sunlight, plus upturned earth and dead branches on the ground, promote growth for plant species and shelter for a variety of small animals. Overall, when African elephants disappear, humans will miss them and the diversity of plants and animals these elephants help provide.

BIBLIOGRAPHY

Allman, W. F., and J. Schrof. October 2, 1989. Can they be saved? (Kenya's effort to conserve its elephant population). U. S. News and World Report 107(13):52-58.

Contreras, J. November 18, 1991. The killing fields: officials in southern Africa say they must shoot elephants to protect them. Newsweek 118(21):86-88.

Merritt, J. April-May 1991. Requiem for a heavyweight: we're learning the ways of elephants while ivory hunters love them to death. Modern Maturity 34(2):56,58-61.

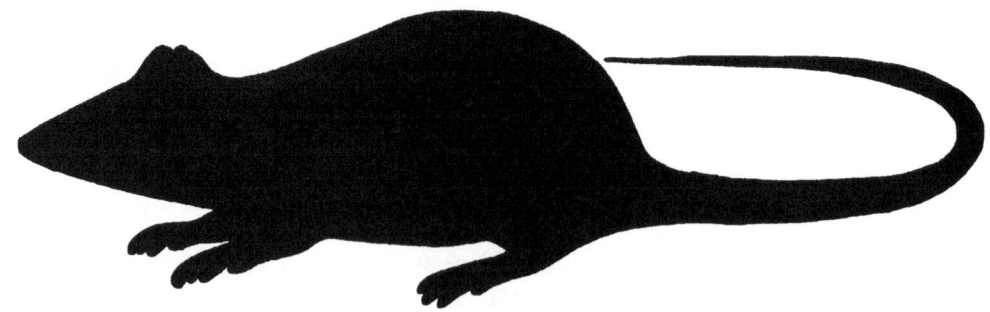

Tony
Gerrard
-94-

COMMENSAL RATS AND MICE

A mong several species of rats and mice in the United States, only three are commensal. Commensal means that these rodents live in close relationship with humans, and they depend on the dwellings of humans as well as their food and waste products. Such rodents include the house mouse *(Mus musculus)*, black rat *(Rattus rattus)*, and Norway rat *(Rattus norvegicus)*. These rodents are the ones frequently used as cartoon characters. Recall Mickey and Minnie Mouse. And, how about the bad character, the one drawn in a trench coat with collar up, Humphrey Bogard hat with front brim down, cigarette in mouth, and referred to as—the rat. In any event, humans introduced all three species into the U.S.

Because of recent arrivals in the U.S., commensals have a different status than native rodents. The latter, which includes such species as voles from northern cold areas, deer mice from wooded areas, kangaroo rats from the desert, and pack rats from several regions, live primarily off native plants and seeds and have no survival relationship with humans. Native rodents originated in the U.S. or arrived there many thousands of years ago by means of periodic land bridges, moving continents, or both. In fact, native mice and rats entered the North American continent—U.S.A.—long before humans.

Commensals arrived much later in the U.S. via ships bringing people and their products. The house mouse came from an area extending from the Mediterranean region to China. There is no official record of arrival time for the house mouse, but it probably occurred sometime in the 1600's. The black rat is from the Malaysian region, and it also arrived in the 1600's. The Norway rat is from China, and it placed its little feet on U.S. shores in 1775. Once on land, all three species quickly associated themselves with construction of buildings and food production as these activities spread across the continent. At the present time, the house mouse and Norway rat live in areas of human habitation everywhere in the U.S. The black rat is restricted to the southern states, especially near coastal areas. But, wherever those species reside, they are not welcome.

All three commensal species have been troublemakers. The rat has been blamed for an estimated $500 million to $1 billion of economic loss each year for the destruction or damage to crops and other food products. Rats kill domesticated animals and wildlife, and they are known to attack about 14,000 persons annually in the U.S. Many diseases are spread by commensal mice and rats, as well as food poisoning. Overall, it would seem there is little good to say about these commensal pests, but where credit is due, it must be given.

Certain strains of commensal mice and rats have been used for experiments in medicine and as pets. These include the white mouse which is a genetically-selected strain from the house mouse and the white rat which is a selected variety of the Norway rat. I am the first to admit that these selected strains can be a delight as a pet. My pet white Norway rat was named Herky, short for Hercules, and I enjoyed his companionship for three years until the time of his death.

BIBLIOGRAPHY

Nowak, R. M. 1991. Walker's mammals of the world. 5th ed. Vol. 2. Baltimore: The Johns Hopkins University Press.

FIRE

T his may come as a surprise to many people, but fires are as natural in nature as are sunshine, water, and wind. The natural origin of fires is lightning, although other sources exist, such as active volcanism. Generally, humans fear fire and consider it far from natural. Perhaps this attitude stems from the loss of personal property, which is understandable. But, to appreciate the overall importance of fire in nature, fire must not be considered on a personal basis. The concern here is the burning of vegetated areas or ecosystems such as grasslands, shrublands, woodlands, and forests in areas where personal property and human resources will not be endangered. In those ecosystems, fires have several natural functions.

Fire releases calcium, phosphorus, and other nonflammable minerals (or nutrients) from organisms and allows these materials to recycle through tissues of other organisms. When live organisms are burned, the minerals become part of their charred remains. Minerals in these remains return to the soil and are used by plants to construct their own tissues. In turn, animals obtain minerals from plant tissues when these plant tissues are consumed.

Fires maintain the existence of grasslands. When grasslands burn, only invading shrubs and trees are killed. Grasses are not killed because their growth points are situated close to the soil's surface; therefore, these growth points escape fire damage. Grasses germinate soon after a fire. This explains why American Indians may have set fires on grasslands to maintain a food base for buffalo and other kinds of grazing animals.

Another function of fires is that they maintain a variety of vegetative patterns in forests. For example, an area subjected to a massive forest fire is devoid of life, but only for a short period of time. The nutrients released through burning act as fertilizer and support a dense growth of herbaceous plants and grasses. Years later, the area is invaded by various species of shrubs. Eventually, various kinds of trees germinate in the rich fire soil and replace the shrubs. Successional stages of tree species continue until a final stage of forest growth is established. This final stage is called the climax because it is this stage that is in balance with the area's climate. In Yellowstone National Park, the development of a climax following a massive fire takes about 300 years.

Other events accompany vegetative changes in a fire area. As succession proceeds, a woody fuel supply composed of trees, shrubs, and deadwood accumulates in the understory of the forest. As the forest gradually matures toward climax, the chances increase for a hotter and hotter fire. Each one of these successional stages in a forest brings forth different species of wildlife. Therefore, fires started over a broad expanse of forest at different times would bring about a variety of successional stages, each with its own assemblage of wildlife. The outcome is wildlife diversity.

Both wildlife diversity and reduction of fuel loads are good reasons for current policies on controlled burnings in forests. Diversity is attained by allowing lightning fires to burn. Usually fires burn a limited number of acres before they extinguish themselves. Fuel load reductions are carried out by prescribed burning programs designed to burn forest understory. These fuel load reductions prevent fires, regardless of origin, from spreading in an uncontrollable fashion.

For many years the motto—prevent all forest fires—was in effect and Smokey the Bear was the symbol for that effort. Unfortunately, this motto may have caused the general public to develop an attitude that all fires are "bad." The recent trend, however, is to control the "bad" fires, the ones started by unauthorized humans and favor the "good" fires, those ignited by lightning or by trained personnel who follow prescribed fire management plans.

BIBLIOGRAPHY

Romme, W. H., and D. G. Despain. 1989. The Yellowstone fires. Scientific American 261:37-46.

Stern, K. R. 1988. Introductory plant biology. 4th ed. Dubuque, IA: Wm. C. Brown Publishers.

REPRODUCTION IN
KANGAROOS

If there were a prize for unique reproductive strategies, the female red kangaroo would likely receive it. This animal's reproductive techniques have been studied quite extensively, and the method used by the red kangaroo to bear offspring is similar to methods used by other species of kangaroos. The red kangaroo can give birth to about three offspring over a period of two years; whereas, most other nonmarsupial grassland grazers, such as African zebras and antelopes, give birth to only one newborn in that amount of time.

The female red kangaroo's reproductive methods are a good example of this unique event. During her first pregnancy, the developing embryo within the oviduct moves to the uterus. Once there, the embryo embeds in uterine tissues and receives maternal nourishment for about 30 days. At that time, while still an embryo and unrecognizable as a kangaroo, the developing individual departs from the birth canal and climbs, unassisted, up over the mother's belly to her pouch. At this time, the mother is in an inclined belly-up position. The embryo, now referred to as a joey, enters the pouch and makes permanent contact with a nipple and begins sucking milk.

As soon as the joey is stabilized on the nipple, the mother mates again. This time, however, the embryo only develops into a 100-cell stage and then becomes dormant in the uterus. This dormancy lasts for six and one-half months which is the length of time the joey is developing in the pouch.

The departure of the joey from the pouch at the end of the six and one-half month period triggers a "go ahead" to the dormant embryo in the uterus. Development of the embryo is initiated and continues for 30 days. At that time, birth begins. The embryo crawls to the pouch and locates a nipple, but not the one used by the joey which just departed from the pouch. That nipple remains the property of the previous joey, now called a young-on-foot, who climbs back into the pouch at mealtime. The young-on-foot continues to use the pouch and its personal nipple for six more months. Apparently milk from that nipple is different from the one being used by the joey who is the young-on-foot's pouch mate.

Meanwhile, the birth of the 30-day-old embryo initiated a mating behavior followed by another dormant embryo. Therefore, reproductively speaking, the female red kangaroo is in the following state of affairs. An embryo is dormant in the uterus, a joey is conducting an uninterrupted nursing process in the pouch, and a young-on-foot climbs in and out of the pouch at will. Finally, the young-on-foot gives up the nipple and leaves the pouch permanently, the joey soon becomes the young-on-foot, another embryo enters the pouch, and another mating takes place—assembly line. The red kangaroo takes the prize for distinctive means of reproduction.

There are definite advantages to this reproductive style. If the young-on-foot should not survive, the joey is a replacement offspring. If the joey perishes, then the embryo is another replacement. Also, this strategy eliminates crowding in the pouch and allows only one developing kangaroo at a nursing station at one time. Finally, and most importantly, this strategy eliminates competition among offspring because they are all different ages.

BIBLIOGRAPHY

Dawson, T. J. 1980. Kangaroos, vertebrate adaptations, readings from the Scientific American. San Francisco: W. H. Freeman and Company.

FACTS AND FABLES
ABOUT SNAKES

So you saw a snake do something, or someone told you they saw a snake do something, and yet, authorities may say the snake did not do what you saw or heard. Now what? Here are some choices—feel confident about what you saw, ponder what someone else said they saw, and keep a mind open to the advice of an authority. At least that is the best way to deal with facts and fables about snakes. Here are some examples of fables.

Some say a hoop snake is able to take its tail into its mouth and roll over the ground like a hoop. From that position, the snake can launch itself onto its enemy. A friend of mine from Oklahoma said that, as a young girl, she actually observed a snake rolling along the ground like a hoop. I was well-aware that I could not convince her otherwise, but despite reports from individuals who say they have seen snakes in hoop positions, experts say such behavior among snakes has never been documented.

One often hears that a venomous snake must be coiled to strike. Authorities indicate that this is not the case. A snake stretched across a trail can turn its head and strike any animal that steps on it or near it. It is true that a snake can strike with greatest force from a coiled position. In fact, coiled rattlesnakes can strike one-third to one-half of their body length.

Rattlesnakes swallow their young during times of danger. This belief may have originated because individuals have observed baby rattlesnakes crawling out of a snake's body from a wound caused by the sharp edge of a hoe or some other sharp object in the hands of a fearful human. These dispersing baby snakes were not swallowed. They are simply the offspring of a female snake. Baby rattlesnakes hatch from eggs inside the body of the mother. Later, during their birth, these snakes exit the mother's body through an opening at the base of her tail called the cloacal vent. In other words, baby rattlesnakes do not hatch from eggs laid outside the body as is the case for many other kinds of snakes.

Some folks say snakes milk cows. A possible reason for this belief is that snakes lay their eggs in compost or manure piles, and when these eggs are discovered, they are sometimes broken in haste. The milky colored albumen which oozes from these broken eggs may be mistaken for milk.

Many outdoorsmen have stated that snakes will not cross a horsehair rope. This fable has caused many a rope-encircled camper to feel quite secure while in a deep sleep. Good luck! because there is no scientific evidence to support that snakes fear ropes. In fact, a rattlesnake encircled with a rope will easily crawl over it to escape.

Range riders of the American Southwest have often heard that injured rattlesnakes will not die until sundown. My personal experiences have proved this story to be false. During my collections of rattlesnakes for biological research, I have observed many injured rattlesnakes die before sundown.

It is often believed that a rattlesnake skin around the neck will cure a cold, a skin around the head will cure a headache or toothache, and rattlesnake fat applied as a lotion will cure rheumatism. Such remedies are not founded on fact or truth.

Certainly stories about snakes are more numerous than those stated here. For many of those stories, there are no written comments which support or refute them. The real answer about validity will likely remain unknown.

BIBLIOGRAPHY

Boys, F., M.D., and H. M. Smith, Ph.D. 1959. Poisonous amphibians and reptiles: recognition and bite treatment. Springfield, IL: Charles C. Thomas.

Shaw, C. E., and S. Campbell. 1974. Snakes of the American west. New York: Alfred A. Knopf, Inc.

Tony Gennaro
95

CARIBOU AND REINDEER

This may come as a surprise, but caribou from America and reindeer from Eurasia are the same species with the scientific name, *Rangifer tarandus*. This really means that there is no difference between Rudolph the Red-nosed Reindeer and Rudolph the Red-nosed Caribou. They both have a very shiny nose because they are one and the same kind of animal. However, the name caribou is still used for members of the species in northern North America and reindeer is commonly used for members in northern Eurasia.

Rangifer from these northern areas differs in a few ways from other members of the deer family. Being dwellers of a cold climate, *Rangifer* is well-endowed with a very thick-haired coat. These deer also have broad, flat hooves which provide an effective means to walk on soft ground and snow, and unlike other deer, both male and female *Rangifer* have antlers.

The features *Rangifer* share with other deer include the shedding of antlers annually, the birth of young in the spring, and a precocial habit. Precocial means that newborns are able to get on their feet and move about quickly after birth. In fact, one hour following birth, these young deer are able to follow their mother, and after one day of life, *Rangifer* can outrun a human. A precocial habit is a must for animals which graze in herds and are consistently on the move to find edible plants.

The requirement for edible food brings about extensive migrations among *Rangifer*. Northern populations migrate from their summer ranges on the tundra to the winter ranges in timbered areas. For example, caribou tracked in Alaska and the Yukon during migration demonstrated that they had annual movements of up to 5,055 km (about 3,000 miles), a record for a migratory land animal.

Unfortunately, many migratory routes of *Rangifer* are being blocked by human activities. These activities include gas pipelines, ice breaking equipment, transportation corridors, and other environmental modifications.

Interruptions of seasonal movements of *Rangifer*, as well as the use of these deer as resources for meat, skins, and antlers have caused *Rangifer* to decline in numbers, especially in Eurasia. In North America, caribou populations have decreased from three and one-half million to about three million. Counts of reindeer in Eurasia have decreased from five million to one million. Overall, world populations of wild *Rangifer* have decreased from eight and one-half million to a count of four million.

These counts do not include the three million domesticated members of *Rangifer tarandus*. This deer was domesticated about 3,000 years ago in Eurasia. Worldwide, domesticated *Rangifer* is a resource for antlers, meat, and hides. Hopefully, in years to come, we will continue to benefit from domesticated *Rangifer,* and at the same time, be fortunate enough to view large numbers of these deer in the wild.

BIBLIOGRAPHY

Nowak, R. M. 1991. Walker's mammals of the world. 5th ed. Vol. 2. Baltimore: The Johns Hopkins University Press.

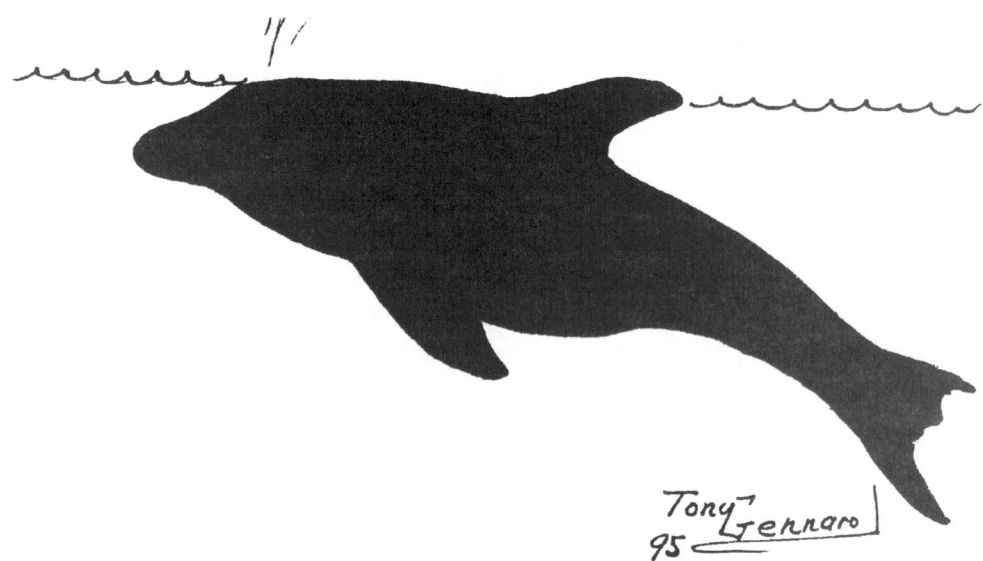

Tony Gennaro
95

AQUATIC ADAPTATIONS OF CETACEANS

Dolphins, porpoises, and whales, otherwise known as cetaceans, are not fishes. Cetaceans are air-breathing mammals. These aquatic mammals entered the sea many thousands of years ago. Since then, several adjustments in the lifestyles of these ceta-ceans have taken place. For example, cetacean nostrils, called blowholes, open at the highest point on the head. Some cetaceans have one blowhole; others have two. Fresh air enters these blowholes, and from there, by way of bronchial tubes, air enters the lungs. A blowhole positioned on top of the head is important because it enables cetaceans to inhale air during their up and down swimming movements. Also, cetaceans can breathe air with an emerged blowhole at the same time they perceive an underwater world with submerged eyes and ears. In other words, cetaceans are expert snorkelers which is important for them because their primary resources for existence (food and mates) are beneath the surface of the sea.

Despite their snorkeling capabilities, cetaceans can remain com-pletely submerged for long periods of time. The sperm whale can remain underwater for 70 minutes, and dolphins can remain sub-merged for about 5 minutes. As a comparison, humans can remain submerged underwater for about one minute.

There are several reasons why cetaceans can remain underwater for long periods of time. One is that cetacean tissues do not become depleted of oxygen as quickly as tissues of land-dwellers. Oxygen is important for body functions because it is associated with energy transfer or storage. Although cetaceans and land animals produce energy in the absence of oxygen, land animals, unlike cetaceans, lack tolerance for lactic acid which accumulates during oxygen-free energy production. Also, land-dwellers cannot tolerate as much carbon dioxide in their tissues as cetaceans. When carbon dioxide accumulates in tissues of land-dwellers, it signals an urge for the land animal to expand its lungs and take in a breath of air. That is not a good idea underwater. Cetaceans can absorb more oxygen per square inch of lung tissue than land-dwellers because cetaceans have a greater number of blood vessels in their lungs. In addition, cetaceans have twice as many oxygen-carrying red blood cells, and two to nine times as much of the oxygen-storing protein, myoglobin, in their muscle as noncetacean mammals.

Cetaceans are unique in their ability to give birth to an air-breathing calf underwater. As soon as the young cetacean passes through the birth canal, it quickly ascends to the surface of the water, an activity aided by the mother in some cases. Certainly, instinctive behavior initiates air-breathing in the newborn calf once it reaches the surface. To initiate fluid-feeding of the youngster, the mother lies on her side on the surface of the water and nurses the calf by means of teats situated in slits on either side of her reproductive opening. The mother contracts her body muscles to force milk through the teats into the mouth of the calf. This is necessary because cetaceans lack lips which enable them to suck. Later, the calf is able to nurse underwater. These adaptations which facilitate the birth of cetaceans are among many adaptations that enable all 79 cetacean species to survive successfully in an aquatic environment.

BIBLIOGRAPHY

Burton, R., M.A. 1980. The life and death of whales. 2nd ed. New York: Universe Books.

Coffey, D. J. 1977. Dolphins, whales and porpoises: an encyclopedia of sea mammals. New York: Macmillan Publishing Company, Inc.

Gaskin, D. E. 1972. Whales, dolphins and seals: with special reference to the New Zealand region. Auckland, New Zealand: Heinemann Educational Books.

Martin, K. 1988. Giants of the sea. New York: Gallery Books.

Martin, R. M. 1977. Mammals of the oceans. New York: G. P. Putnam's Sons.

SHARKS AND BONY FISHES, A FEW DIFFERENCES

Mention the word "shark" and one's imagination strays from that of a gentle, docile fish. What comes to mind is a slow-moving, grayish, streamlined creature with a pointed snout, triangular teeth, slits on the sides of the body, a large fin on top of the body, and a long, pointed tail fin. They certainly look like fish, but is a shark a fish? The word, shark, has commonly been used separately from the word fish for a long time. The answer is yes. Sharks are fishes—all 339 species of them.

Shark features are fish-like, but there are a few distinct differences between sharks and other fishes. For example, sharks have a carti-laginous skeleton; most other fishes have a bony skeleton, hence the name, bony fishes. Unlike bony fishes, sharks lack a swim bladder, and their gill breathing-mechanism is different from bony fishes. In fact, the shark's gill mechanism has led to the belief that they must swim continuously to breathe, and there is truth to that idea.

Most sharks must swim to breathe, except for catshark, nurse, carpet, and horn sharks. These four species of sharks can remain stationary on the floor of the ocean and breathe very effectively because of a certain feature associated with their gill breathing-mechanism. Bony fishes also have this feature, and they can remain motionless at various depths and breathe. The feature is better understood once the breathing-mechanism is explained.

The mechanism works in this manner. Water enters the mouth of fishes and flows over gills before it leaves their body through openings on the side of the head. As water flows over the gills, oxygen within it diffuses into the gills and enters blood vessels, where oxygen is used in metabolic functions. The amount of oxygen required for metabolism demands a continuous inflow of water. To maintain flow, fishes must do one of two things—initiate forward motion to allow water to enter the mouth or pump water into the mouth. Most all bony fishes and sharks, such as catsharks, nurse, carpet, and horn sharks pump water into their mouth; therefore, they can remain motionless and breathe. Other species of sharks cannot pump; they must swim.

Fortunately, sharks accomplish another feat when they swim. They do not sink. Although some buoyancy is provided to them by their livers which are heavily laden with oil, sharks seem not to have the buoyancy of bony fishes.

Most bony fishes easily remain afloat because of their air bladders. The bladder is gas-filled, whitish, thin-walled, and situated inside the fish's body. To reach greater depths in the water, bony fishes remove gases from the bladder. To reduce depths, they add gases to the bladder. In some species this exchange of gases is between the bladder and blood. In others, the exchange is between the bladder and atmospheric air. These species gulp air at the surface of the water to fill the bladder and remain near the surface, or they burp air to descend to lower depths.

Ultimately, air bladders are an advantage to bony fishes in several ways, depending on the species. Keeping afloat with ease is one. This floating feature also restricts some species to certain ranges of depth in the water. Restrictions of similar species to different levels in the water reduces competition for resources among these similar species.

It is a definite advantage to sharks that they are not restricted to specific levels in the water because they lack an air bladder. Sharks are predators, and it is profitable for them to exploit prey resources

from all levels of their habitat. That may be why their prey prefer-
ences are broad, including other fishes, sharks, seals, sea turtles,
crabs, lobsters, and shrimp.

BIBLIOGRAPHY

Castro, J. I. 1983. The sharks of North American waters. College
    Station, TX: Texas A&M University Press.

Cousteau, J., and P. Cousteau. 1970. The shark: splendid savage
    of the sea. Garden City, NY: Doubleday and Company, Inc.

Dennis, F. Editor. 1975. Man-eating sharks: a terrifying compilation
    of shark-attacks, shark-facts, and shark-legends! Secaucus, NJ:
    Castle Books.

Ellis, R. 1976. The book of sharks. New York: Grosset and Dunlap
    Publishers.

Lineaweaver, T. H. III, and R. H. Backus. 1970. The natural history
    of sharks. Philadelphia: J. B. Lippincott Company.

McFarland, W. F., F. H. Pough, T. J. Cade, and J. B. Heiser. 1979.
    Vertebrate life. New York: Macmillan Publishing   Company.

Scharp, H., and M. Scharp. 1976. Shark wanted! dead or alive.
    South Brunswick, NJ: A. S. Barnes and Company, Inc.

MEDICINAL VALUE OF PLANTS

The pills and other medications we take for our health which come in different sizes and colors are not produced entirely from synthetic materials in the laboratory. In fact, one-fourth of those pills contain at least one natural product extracted from a plant. It is likely that additional plants would be used for medicinal purposes if we knew more about them. Chances to obtain such knowledge are becoming less and less likely. The reason is that demands from an increasing global population of humans are initiating a gradual disappearance of tropical rain forests. These forests contain one-half the number of species of plants on Earth.

As rain forests gradually dwindle in size, so do the numbers of primitive human cultures which inhabit them. This reduction of cultures is caused by the disappearance of primitive human habitat, as well as the spread of modern technology into those cultures. Various primitive people have exploited plants for medicinal purposes for centuries, but unfortunately, written records of their uses were not established. Such information was passed down verbally from one generation to the next. Therefore, it is virtually impossible for scientists to recapture those medicinal practices. But all is not lost—yet.

Two primary modes of action will enable scientists to acquire knowledge about medicinal plants. One is to lobby for means which will decrease the rate of destruction of tropical rain forests. This decrease may give workers time to examine plant species for their medicinal value. The other is to collect information about medicinal plants from the remaining elders of primitive cultures. Either action is a difficult and lengthy task, but the effort is enhanced by knowledge already known about some medicinal plants.

Here are some examples of medications from plants we already know about. Drugs extracted from the periwinkle of Madagascar are used to treat leukemia and Hodgkin's disease. Drugs obtained from the pareira of South America are important muscle relaxants. Contraceptive pills and cortisone are produced from yams grown wild in Mexico. These are all important curing agents, but medicinal plants are not restricted to use by humans.

Wildlife also uses plants for medicinal purposes. Chimpanzees swallow certain leaves whole. These plants have oils that apparently kill disease-causing bacteria, fungi, and parasitic worms in the digestive tract. One researcher observed the diet of a pregnant wild elephant for a year. At the end of that year, the elephant took an unusually long trip to a riverbank and consumed an entire tree. Four days later, she gave birth to a healthy baby elephant. That kind of tree was never part of her diet. Interestingly, humans use the same remedy, in that pregnant women in Kenya commonly brew a tea from the bark and leaves of the same tree to induce birth of their children. Rhesus monkeys eat dirt from certain areas because it contains a high concentration of the active ingredients in Kaopectate which is used to treat diarrhea.

These are just a few remedies witnessed among wildlife. One wonders if humans will ever discover the many other medicinal uses of plants used by wildlife worldwide, not just from forests, but from deserts and grasslands as well.

BIBLIOGRAPHY

Krochmal, A. 1973. A guide to the medicinal plants of the United States. New York: Quadrangle.

Lewis, W. H. 1977. Medical botany: plants affecting man's health. New York: John Wiley and Sons.

Stolzenburg, W. 1990. Garlic medicine: cures in cloves? Science News 138:157.

Tony Gennaro
94

RATTLES OF RATTLESNAKES

Hearing a series of sharp, short sounds from the rattles of a snake on television, radio, or from a movie soundtrack is one thing, but actually hearing them in the wild is a completely different story. The sound resembles water hitting cement after a hydrant has been suddenly turned on with full force.

The sound is produced when a series of loosely fitted, hollow rattles at the end of the snake's tail contact one another repeatedly as the tail vibrates. The very thin walls and hollowness of each of the rattles contribute to its sound. Unlike baby rattles, the sounds are not made from particles hitting the sides of the rattle when the rattle is vibrated, and snake rattles have a more complicated structure.

The development of snake rattles is unique. A rattlesnake's skin is composed of scales. Each time a rattlesnake molts, the entire skin is shed, except for one modified hollow scale at the end of its tail. The sequence of rattle formation begins with the first molt. Following that molt, the modified scale remaining at the very tip of the tail is called a button. A young rattlesnake with only a button cannot rattle. It requires two or more rattles to make the rattling sound. When the snake molts a second time, another hollow scale remains loosely attached to the outside of the initial button. The next molt results in a third hollow scale attached to the second. Through this process, a new hollow rattle is added with each molt, forming a linear series of hollow scales loosely attached to one another.

The number of rattles does not indicate the age of the rattlesnake because molting occurs more than once each year. In fact, snakes may undergo two to four molts annually. The number of molts per year varies among species of snakes, but regardless of the frequency, not all of the rattles remain attached. Only about ten rattles are retained on a rattlesnake; the remainder break off. If all were to break off, the snake could not give a warning, and it could be a serious problem for the snake and the intruder.

Rattlesnakes rattle to protect themselves or capture prey. For example, rattlesnakes rattle to avoid being stepped-on by grazing animals which are great threats. Rattlesnakes have spent thousands of years in the same habitat with buffalo, pronghorn, horses, and many other grazers. A coiled snake resembles a pile of dung deposited on the prairie, but a warning rattle informs grazers that the snake is not dung. On the other hand, a vibrating, noisy tail may serve as a distraction in prey capture. As the prey concentrates on the vibrating tail at one end of the snake, the prey is zapped by fangs at the other end.

How should a human react to the sound of a rattlesnake in the wild? For those who experience rattlesnakes for the first time, I recommend remaining motionless until the snake is located. Then, simply back off slowly in a direction opposite its location. Those who have encountered rattlesnakes at a previous time may behave differently, depending on their experiences with that snake. As a naturalist who has had at least 50 or more confrontations with rattlesnakes, I have responded in one of three ways: 1) collected the snake, dead or alive, for scientific study, 2) removed and released it in a remote area, free from the activities of humans and domestic animals, or 3) if the area was remote, I simply watched the snake, and invariably, it would crawl off into adjacent vegetation. Rattlesnakes give warnings to larger animals to keep from getting stepped on or killed. If the encroaching animal heeds the warning and backs off, there is little danger to either, and both can continue with their way of life.

BIBLIOGRAPHY

Boys, F., M.D., and H. M. Smith, Ph.D. 1959. Poisonous amphibians and reptiles: recognition and bite treatment. Springfield, IL: Charles C. Thomas.

Burton, R., M.A. 1978. Venomous animals. New York: Crescent Books.

Carr, A. 1963. The reptiles. New York: Time Incorporated and the Editors of Life.

Davidson, T. M., M.D., and S. F. Schafer. April 1989. Rattlesnakes: the animal and the venom. The Physician and Sports Medicine 17(Part 1 of 2).

Gans, C., and R. B. Huey. 1988. Biology of the reptilia: defense and life history. New York: Alan R. Liss, Inc.

Shaw, C. E., and S. Campbell. 1974. Snakes of the American west. New York: Alfred A. Knopf, Inc.

FIRE ANTS

From appearance and size, fire ants, *Solenopsis invicta*, should not be a threat to humans. These red ants are only three-sixteenths of an inch in length, about 575 times shorter and 300,000 times lighter in weight than the average human. Yet, fire ants can cause Jane or John Doe to stand tall, alert, and wide-eyed.

There are good reasons for such reactions. The pinpoint burn from a fire ant sting is not forgotten. Pain from the sting results from higher levels of piperidine alkaloids in fire ants than in other ant species. This venom is injected through a needle-like stinger situated on the end of the ant's abdomen. Unfortunately, the victim usually encounters more than one sting because fire ants typically attack in large numbers.

This painful experience induced by fire ants is relatively new to human inhabitants of the United States. Fire ants were accidentally introduced into the U.S. from Brazil about 50 years ago. At the present time, the introduced fire ant occupies 11 southeastern states from southern Virginia to central Texas.

Since their arrival on North American soils, fire ants have changed their mode of existence. At first, the ants lived in single colonies called monodynes which contained one queen and up to one-half million ants instead of several hundred ants which is common for other ant species. Each monodyne was separated by territorial boundaries and contained a network of tunnels radiating 50 feet outward from the above-ground mound. Recently, however, territorial boundaries have been dissolved, and the colonies have hooked up with one another, forming interconnecting super colonies called polydynes. Polydynes, with their numerous egg-laying queens, have increased the reproductive rate of fire ants tenfold. In some areas of the South, fire ants exist as a thick sheet in the soil, advancing up to a few hundred feet each year. In such areas, one may hop quickly, but carefully, from one anthill to another for a great distance. Counts on one field in Texas, for example, included 500 to 600 above-ground mounds per acre. As the polydynes advance, fire ants destroy life in their path, including other ants, arthropods (ticks, spiders, insects), vegetation, ground-dwelling birds, and even small pets. In fact, counts have indicated that fire ants reduce native ant populations by 70% and arthropods by 40%. The real threat from fire ant attacks is reduction in biodiversity, that is, a reduction in variation of life itself.

Obviously, control of fire ants is necessary; fortunately, planning for this procedure is underway. Experiences from a federal program in 1957 demonstrated that poisons alone will not control numbers of fire ants. Poisons simply kill the native insects and make way for new invasions of fire ants. More recent control programs minimize poisons and maximize educational efforts.

Biological control is the method nature uses to control fire ants in South America where they are native. This kind of control involves restraints by other species on fire ant densities. A control agent in South America, which is the first choice for U.S. biologists, is a protozoan parasite, *Thelohania solenopsae*, which infects the fat and blood cells of fire ants. Ants are weakened by this infection, and the ant colony eventually dies.

Experience with invading fire ants demonstrates the complexity of species interrelationships. In their native habitats, species keep each other in control. However, when a species invades a new area where controls are nonexistent, the invading species goes unchecked.

BIBLIOGRAPHY

Adams, S. 1994. Fighting the fire ant. Agricultural Research 42(1):4-9.

Holldobler, B., and E. O. Wilson. 1990. The ants. Cambridge, MA: The Belknap Press.

Mann, C. C. 1994. Fire ants parlay their queens into a threat to biodiversity. Science 263:1560-1561.

Tony
95 Tennan

PROTECTED REDWOODS

Behold the redwood. Gaze at its massive trunk and follow it upwards to the tip which pierces the sky at about 300 feet, the height of a 30-story building. With head bent back to the maximum, your comment is likely to be, "Wow! Unbelievable!"

There are many people to thank for the continued existence of these magnificent giants. They include individuals in government and in various private organizations. Private organizations have generated millions of dollars to purchase and protect several thousand acres of old-growth redwood forests. These old-growth forests are original virgin forests, as opposed to secondary growth which results from planting saplings after logging operations.

Protected redwood sites are in each of two areas in the United States and one area in the Hubei Province in central China. These three areas are the only places where redwoods survive, although they once covered vast areas of the earth and originated 160 million years ago. The Dawn Redwood, *Metasequoia glyptostroboides*, is native to China. There are about 3,000 Dawn Redwoods remaining. The other two redwood areas are in California. One area is endowed with the Giant Sequoia, *Sequoiadendron giganteum*, which grows in scattered pockets along the western slopes of the Sierra Nevada, a mountain range in eastern California. The Giant Sequoia parks include Calaveras Big Trees Park, Yosemite National Park, Sequoia National Park, and Kings Canyon National Park. Four of the largest trees grow in the Sequoia National Park which includes the General Sherman Tree, the world's largest. The Coast Redwood, *Sequoia sempervirens*, grows in the other area in California which comprises a narrow, almost continuous strip about 450 miles long and 5 to 10 miles wide along the California coast from San Luis Obispo north to the Oregon border. Within that strip, Coast Redwoods are protected within 40 parks, monuments, reserves, and recreational areas.

It is unlikely that protected redwoods will ever become floors on outdoor patios since only a small amount of old-growth, virgin redwood remains in private ownership. Redwood lumber is produced primarily from second-growth forests, and it is not plentiful enough to meet public demands. Since 1965 the volume of redwood lumber has declined, and the value of this precious wood has increased.

BIBLIOGRAPHY

Dewitt, J. B. 1993. California redwood parks and preserves. San Francisco: Save-the-Redwoods League.

Raven, P. H., R. F. Evert, and S. E. Eichhorn. 1992. Biology of plants. 5th ed. New York: Worth Publishers.

Stern, K. R. 1988. Introductory plant biology. 4th ed. Dubuque, IA: Wm. C. Brown Publishers.

POISON-DART FROGS

Humans pay more attention to bright colors. Consider the attention given to red stop signs, yellow curbs, yellow and red colors in McDonald's advertisements, and the display of bright colors which attracts little tots to the toy store in the shopping mall.

Likewise, colors are important to wildlife. Colors attract members of the same species—males attract females, and vice versa. Conversely, colors repel members of the same species—reddish breast of the male robin repels males of the same species.

Bright colors of one species of wildlife may serve also as a warning to other species, especially predators. An example is the array of bright attractive colors among the small (one-half to three inches long) poison-dart frogs of Central and South America. These frogs display one or a combination of the colors red, orange, yellow, and even bright blue which cause poison-dart frogs to be extremely conspicuous against the green vegetation of their habitat.

For this animal, these colors serve as a warning to predators. Fifty-five of the 135 known species of poison-dart frogs secrete a poisonous substance, batrachotoxin, from tiny pores in their skin. This highly toxic substance causes a painful experience or death to individuals, other than members of the same species, that make contact with the frog. For example, one poison-dart species, *Phyllobates terribilis*, has a high enough level of batrachotoxin to kill about 20,000 house mice or several adult humans. The toxin causes irreversible muscle contractions leading to heart failure. Any animal that survives such a contact learns to carefully avoid the frog.

Central and South American Indian hunters use this toxin, which is obtained from three species of poison-dart frogs with high levels of batrachotoxin, to poison the tips of darts. There are at least two ways in which the toxin is obtained. In one case, a pithed frog (frog killed by destroying its brain) is placed over an open fire. The heat from the fire causes toxin to secrete from the frog's skin. The toxin is then applied to a dart. In this procedure, and the one that follows, the frog is handled with a leaf which is used as a barrier against the poison. This includes the handling of *Phyllobates terribilis* which has such high levels of batrachotoxin that enough toxin is obtained by simply moving a dart across the skin of this frog when it is still alive. The toxin from *Phyllobates terribilis* is potent enough to remain active on a dart tip for more than a year. Once prepared, these poison-tipped darts are propelled through the air from a blowgun.

The potency of frog toxin seems to come from diet. Frogs captured in the wild and retained in captivity gradually lose their toxicity, and those reared in captivity are nontoxic. Retention of toxicity is virtually impossible because the diet of insects consumed by dart frogs in the wild cannot be replicated in captivity.

In addition to batrachotoxin, there are many other chemical compounds in the skin secretions of poison-dart frogs which may be of value to humans. One species secretes a substance which is a painkiller 200 times more powerful than morphine. The secretion of another species has a chemical substance which may be useful to reactivate heart-action for patients who have experienced a heart attack. Several other compounds have been isolated from skin secretions of the poison-dart frog and are being analyzed for their chemical and medicinal properties.

Another value of poison-dart frogs to humans is their use by collectors, who rear them for sale or trade. Since wild populations are now protected by international laws, all rearing of poison-dart frogs is from captive specimens.

BIBLIOGRAPHY

Daly, J. W., C. W. Myers, and J. E. Warnick. 1980. Levels of batrachotoxin and lack of sensitivity to its action in poison-dart frogs
(*Phyllobates*). Science 208:1383-1385.

Goodenough, J., B. McGuire, and R. A. Wallace. 1993. Perspec-
tives on animal behavior. New York: John Wiley and Sons, Inc.

Moffett, M. W. 1995. Poison-dart frogs lurid and lethal. National Geographic 187:98-111.

Nemuras, K. T. 1980. The jungle's living jewels. America's 32:25-32.

DOLPHIN COMMUNICATION

The sea is the world of dolphins where they seek prey, escape from predators, and communicate with other dolphins. Interdolphin communication is important because dolphins are herding, flocking animals which behave socially like antelope, cattle, horses, and others which also herd and flock. One dolphin, the bottle-nosed dolphin, *Tursiops truncatus*, forms large aggregations which consist of subgroups of 2 to 15 dolphins.

The concern here is interdolphin communication primarily among bottle-nosed dolphins, a species which is many times the star performer at marine exhibits. It behaves in a similar way to other species of dolphins and depends primarily on sound for environmental perception. Sound, as a means of perception, is in effect day and night.

Sound perception in the bottle-nosed dolphin is accomplished by clicks and whistles. These vocal emissions were once thought to originate from nasal air sacs deep within the structure of the head. More recently, however, researchers demonstrated that clicks originate from tissue near the top of the head close to the blowhole. Apparently, clicks and whistles are not emitted from the larynx.

Bottle-nosed dolphins use clicks to perceive information about the environment. Once emitted from the dolphin, sound waves comprising these clicks contact an object and then bounce back as echoes to sensory tissues of the dolphin. The dolphin then experiences a mental image of the object which indicates its distance, size, shape, and texture. This process of interpreting reflected sound waves is called echolocation.

Whistles, on the other hand, are used to maintain contact with other dolphins. Each dolphin has its own distinctive whistle. In fact, it takes about one second for a dolphin to recognize the whistle of another dolphin. Bottle-nosed dolphins communicate in ways other than the use of clicks and whistles. These marine mammals can see, touch, and taste, but they apparently cannot detect odors.

The subject of dolphin perception and communication generally leads to such questions as, "Do dolphins communicate with one another by a whistle language?" and "Are dolphins capable of a language communication with humans?" As stated previously, dolphins communicate with one another, but this appears to involve an exchange of acoustic (sound) signals, not a language. Other animals convey acoustic signals by their barks, growls, chirps, or whatever sounds they emit. Also, most authorities conclude that dolphins cannot communicate with humans by a language. Attempts by humans to communicate with dolphins through words and whistles have been attempted, but thus far, results are less than exciting. A language consists of the ability to put symbols together according to a system which brings about a more or less unlimited number of combinations which convey intended messages.

The dolphin's agility, keen perception, imitative behavior, ability to learn quickly, and large brain suggest it possesses a high level of intelligence. However, current evidence suggests that the large dolphin brain is associated mainly with hearing and the processing of echolocation and other communicative data. The dolphin brain appears not to house a center for high intelligence. Therefore, those individuals who are still willing to wager that dolphins are as smart or smarter than humans will have to await further research findings to collect or lose their bets.

BIBLIOGRAPHY

Caldwell, M. C., and D. K. Caldwell. 1979. Communication in Atlantic bottle-nosed dolphins. Sea Frontiers 25:13-138.

Cousteau, J., and P. Diole. 1975. Dolphins. New York: A & W Publishers, Inc.

Nowak, R. M. 1991. Walker's mammals of the world. 5th ed. Vol. 2. Baltimore: The Johns Hopkins University Press.

Parfit, M. 1980. Are dolphins trying to say something, or is it all much ado about nothing? Smithsonian 11:73-81.

Peterson, I. 1992. Dolphin sonar: using their heads to click. Science News 142:325.

Wursig, B. 1979. Dolphins. Scientific American 240:136-148.

Tony Gennaro
95

MIMICRY

Visualize this situation. You are walking in a wooded area, and a tiny flying insect with orange and black colors buzzes your head. You shoo the creature away, but it continues to return. Then, the insect lands on your arm. You give the pest a fast sweep with the hand. The insect falls to the ground belly up. You assume you have saved yourself from a bee sting. That is possible, but you may have zapped a fly. Many species of flies are in bee's clothing. Why? Well, this is one of nature's many tricks.

This imposter game is known as Batesian mimicry, named after an English naturalist, Henry W. Bates. In 1862, Bates published his interpretation of why different species of butterflies (some edible to predators, some not) are similar in appearance. He stated that mimicry is a process in which an edible species of butterfly, the mimic, resembles an inedible species of butterfly, the model, in order to deceive a predator. This process gives the edible species the same chance of survival as the inedible species because any predator which experiences a bad-tasting butterfly will avoid another encounter. In the aforementioned example, the pesky orange and black insect could have been a fly which resembled a bee in color. Had this fly come in contact with a hungry bird or moth, it is likely that the mimic fly would have escaped untouched if the bird or moth had previously experienced the painful sting of a bee.

Since Bates' publication, several other examples of mimicry have been reported. Mimics may communicate their "copy-cat" features visually, such as the fly and butterfly mimics, or they may mimic in other ways, or a combination of ways. For example, species may sound, behave, or emit odors similar to the species they mimic. In other words, a harmless insect may give off the same odors as a noxious one, and a predator will leave them both alone.

Here is an example in which the mimic is allowed access to space occupied by the model. The mimic feature used is visual. A cleaner fish, the sea swallow, is territorial and inhabits coral reefs in the Pacific and Indian oceans. Other fish species enter the sea swallow territory to be cleaned, that is to have parasites and fungal skin growths removed. Sea swallows and the sabre-toothed blenny, an aggressive flesh-eater, are almost identical in color. Sometimes customer fishes approach the sabre-toothed blenny. Instead of being cleaned, customers are attacked by the sabre-toothed blenny, and pieces of the customer's flesh are torn away and eaten.

Another example allows entrance to a model's nest (hive or colony). Although the environment of such nests is hostile to invaders, visual and sometimes odoriferous mimic features are effective and the mimic is not detected. Nest mimics include species of flies and moths which resemble and enter nests of bees. Likewise, species of spiders and beetles resemble ants and enter their nests. Once inside, mimics scavenge available materials.

One very effective way to survive is to mimic a predator both visually and behaviorally. Mimicry of this kind is accomplished by the snowberry fly which mimics in appearance its predator, the zebra spider. When the snowberry fly is disturbed, it executes a characteristic defensive display. Wings, with banding patterns which resemble the legs of a spider, are brought slightly forward by the fly and moved in a way to resemble the gait of a zebra spider. This performance fools the zebra spider and extends the life of the snowberry fly. Because of its usefulness, mimicry is firmly carved into the natural mechanisms for survival.

BIBLIOGRAPHY

Mather, M. H., and B. D. Roitberg. 1987. A sheep in wolf's clothing: tephritid flies mimic spider predators. Science 236:308-309.

Owen, D. 1980. Survival in the wild: camouflage and mimicry. London: The Oxford Press.

Pough, F. H. 1988. Mimicry of vertebrates: are the rules different? Pages 67-95 *in* L. P. Brower's Mimicry and the evolutionary process. University of Chicago. American Naturalist 131:67-95.

Neat Things About Skunks

A friend of mine has great respect for nature. He once asked me, "Tell me something interesting about skunks." I paused for a moment and responded with, "Robert, I can think of three neat things about skunks." I began with their coloration.

The coloration of skunks is designed to advertise rather than conceal. Nature's purpose for skunk coloration is to teach an assailant to associate its nasal irritation from skunk spray with the color of the skunk. The desired result is that the assailant will not pursue the skunk again. But, why black and white and not green and red, or yellow and purple?

Research on humans may provide an answer. Certain cells in our eyes, similar to the cells of other vertebrates, respond strongly to the contrast between the edge of two colors. What greater contrast is available than the junction of the darkest color and the lightest color as found on a skunk? If the skunk were pure black or pure white, it could not advertise as well. As an example, a black mouse hides on lava rock, and a white snowshoe hare hides on snow.

What is the source of the skunk's odoriferous spray? This fluid or musk is secreted from a pair of glands adjacent to the anus. In the striped skunk and hooded skunk, the musk is discharged either as a spray or as a short stream of rain-sized drops. When either of these skunks is confronted by an antagonist, the skunk arches its back, elevates the tail, erects its hairs, and stamps its feet on the ground. Finally, the skunk turns the body in a U-shaped position, with the head and tail facing the intruder, and discharges the fluid. This fluid usually travels two to three yards, but the smell can be detected downwind up to one and one-half miles. One squirt is enough to send the most ferocious, snarling, salivating canine yelping in agony.

Another interesting fact about the skunk is an unusual function in reproduction called delayed implantation. This means that the fertilized egg does not implant immediately when it enters the uterus. In the western spotted skunk, the developing embryo enters the uterus from the oviduct and floats freely for 180 days. Then, the embryo implants, develops for 30 days, and the young are born in the spring. This would imply a pregnancy of 210 days, but the actual time of embryonic development is only 30 days. One can only guess the adaptive significance of delayed implantation. Perhaps the process allows for favorable environmental conditions for a fall copulation and a spring birth, both of which are on either side of a long, hard, cold winter.

A little-known idea about skunks is that they may feed on rattlesnakes extensively. This event is suspected because skunk musk causes an alarm reaction in rattlesnakes, and both spotted skunks and hog-nosed skunks are resistant to the venom of rattlesnakes.

BIBLIOGRAPHY

Alcock, J. 1993. Animal behavior. 5th ed. Sunderland, MA: Sinauer Associates, Inc.

Nowak, R. M. 1991. Walker's mammals of the world. 5th ed. Vol. 2. Baltimore: The Johns Hopkins University Press.

Vaughan, T. A. 1986. Mammalogy. 3rd ed. Philadelphia: W. B. Saunders Company.

ANIMAL PLAY

We don't usually associate playing with birds and mammals, but it is very common and they seem to relish it. There is play fighting (animals scuffling without anger), object play (one animal playing with an object), social play with objects (several animals playing with objects, such as sticks or dead prey), and complex social play (animals making toys of objects and engaging in creative behavior). Does this play have function? Some authorities say no; others say yes.

Those that say yes state that play develops social relationships and status among members of a species and teaches animals about their environment. Furthermore, these experts conclude that play improves skills for fighting, hunting, and mating. Those who support play as functional in life state that young, playful animals learn to be flexible, inventive, and adaptive.

These comments are based primarily on observations and logic rather than direct evidence because experiments to test play function are difficult to design; however, other scientific findings about play have emerged. Researchers have shown that play varies with environmental conditions. For example, among young gelada baboons, abundant food during the rainy season allows more time for play than during the dry season when food is scarce. Also, among a certain species of the rat, investigators have demonstrated that play varies with time of weaning. Early rejection of sucklings resulting from food-starved mothers causes these young rats to initiate play faster than when they continue to nurse. This early play apparently accelerates social development and prepares the young rats for an independent social life. Another study demonstrated that play may be a high risk among South American fur seal pups. The author assumed that if fur seals continue to play even though play may cost their lives, then play must be beneficial. The study indicated that while playing, 22 (85%) of 26 fur seal pups were killed by sea lion predators. Despite this kill rate, play persisted. The author concluded that costly play must be beneficial to the species of fur seals in the long run.

So what about young humans? Do they require play to experience a normal adult life? Stuart L. Brown, a former physician psychiatrist who studies play behavior, gathered some interesting facts. He based his observations on data from 26 convicted murderers. Brown showed that 90% of the murderers displayed absence of normal play and presence of abnormal play at an early age. Abnormal play consisted of bullying, extreme teasing, or cruelty to humans or animals. Further, Mr. Brown's studies showed that from among 25 drivers (most were intoxicated) who had killed someone else or died themselves in an automobile crash, 75% had childhood play abnormalities. These findings are only correlations. There is no evidence for cause and effect, which means no proof exists that lack of play causes criminal or antisocial behavior. However, such correlations are food for thought. Perhaps healthy, young play provides skills to develop a normal adult social life.

BIBLIOGRAPHY

Barrell, L., R. I. M. Dunbar, and P. Dunbar. 1992. Environmental influences on play behavior in immature gelada baboons. Animal Behavior 44:111-115.

Brown, S. L. 1994. Animals at play. National Geographic 186(6):2-35.

Harcourt, R. 1991. Survivorship costs of play in the South American fur seal. Animal Behavior 42:509-511.

Smith, E. F. S. 1991. The influence of nutrition and postpartum mating on weaning and subsequent play behavior of hooded rats. Animal Behavior 41:513-524.

PIRANHAS

Stick a hand in certain South American waters, and you may pull out five skeletonized fingers. This has been the common belief for waters inhabited with piranhas. This belief may not be entirely true, according to individuals who state that piranhas are not human flesh-eaters as commonly thought. However, discrepancies resulting from reports about feeding behavior of piranhas make it difficult to separate fact from fiction. Further study reveals the true nature of the piranhas.

The first accounts of the piranhas by white men came from the Conquistadors. These freshwater fishes occur only in lakes and rivers in South America. From among several species of piranhas, only four are said to be dangerous to humans. Size ranges from 8 to 20 inches, and the largest piranha rarely weighs more than 3 pounds. Piranhas move in shoals and display sharp, conspicuous teeth. Quick-reacting muscles enable piranhas to open and close their jaws within five milliseconds, faster than the blink of an eye. Prey of piranhas includes small fishes, lizards, frogs, snakes, and on occasion, members of their own species. Roots, leaves, and algae are also on the piranha menu. As a food fish, the piranha is rated second class. However, these fishes are a staple for poor people in some South American areas.

Except for their exposed teeth and reputation of shredding and skeletonizing victims, piranhas are similar to most fishes. So, where did their ferocious, gory reputation come from? One author stated that Theodore Roosevelt returned from a 1913 expedition to South America with numerous accounts from Amazonian storytellers of piranhas attacking humans. Theodore recorded the stories in his journal and initiated the birth of a human-eating piranha.

However, recent visitors to South America strongly disagree with these bloody piranha tales. One renowned fish biologist stated that piranhas never conducted unprovoked attacks. He added that people swim unconcerned and uninjured near areas where piranhas are numerous. Another individual said that for more than 20 years he studied Indian life in South America. From thousands of Indians he met, only seven had been injured by piranhas, and those bites were not serious.

Other individuals admit that the piranha's sharp teeth have been a threat to humans. They comment on fishermen who experienced missing fingers, toes, and chunks of flesh. Other observers indicated that hungry and starving piranhas can be dangerous to injured or weakened people who fall into the water. Furthermore, well-founded reports indicate that piranhas in tributaries of the Rio Paraguay attack when waste from large slaughterhouses or cattle farms are thrown into the water. Perhaps these waters initiate a feeding frenzy, and at that time, piranhas do not distinguish between waste and human flesh. Piranhas about one-half mile away, in water not polluted by blood and flesh, are harmless, and people swim among them. Overall, these comments place the South American fish in a slightly different light than previous reports. Like other animals with flesh-eating capabilities, the piranha's attitude at mealtime reflects body needs and environmental conditions.

Regardless of human concerns about injury, piranhas play an important role in the ecology of South American fishes. Piranhas prey primarily upon wounded and diseased fishes. Therefore, piranhas reduce the spread of fish diseases and fish die-offs. To increase the number of valuable fishes for consumption, poisoning programs for piranhas have been conducted. However, these programs may decrease valuable fish densities because epidemics once controlled by piranhas may well proceed uncontrolled. As a reminder, even fishes with unpleasant reputations (warranted or not) play an important role in the total ecology of their habitat.

BIBLIOGRAPHY

d'Aulaire, P. O., and E. d'Aulaire. May/June 1986. Piranhas.
International Wildlife 16(3):31-35.

Migdalski, E. C., and G. S. Fichter. 1976. The fresh and salt-water
fishes of the world. New York: Alfred A. Knopf, Inc.

Nelson, J. S. 1994. Fishes of the world. 3rd ed. New York: John Wiley
and Sons, Inc.

TARANTULAS OF THE

AMERICAN SOUTHWEST

Picture this—a movie scene showing a large spider crawling along the headboard in a room of an old southwestern hotel. An unshaved, clothed range rider is sprawled on the bed. In walks his partner, who draws a six-shooter and blasts the spider into a cloudy mist. A wasted shot. The "hero" behind the blazing gun just killed a harmless tarantula.

Tarantulas instill fear in many humans. This fear may stem from the large size of these spiders with leg spans of six or seven inches. Although tarantulas have venom, the effect is no more painful than a bee sting. Likewise, the larger tarantulas with leg spans of 10 inches, found in Central and South America, are harmless. The only tarantula to fear is the Sydney funnel-webbed spider, *Atrax robustus*, of Australia which is extremely harmful to humans.

Aside from suspected, yet unwarranted fears, tarantulas have many beneficial attributes. Southwestern tarantulas are insect-eaters which live in burrows in deserts, grasslands, and edges of cultivated fields. In fact, these burrows are tarantula birth-sites. Tarantulas hatch there from eggs deposited by the female following copulation with a mature male. Newly-hatched spiders remain together for a while after hatching, then they disperse from hatching sites and seek a burrow of their own. Both sexes mature at about 10 to 12 years of age. Until that time, males and females are indistinguishable. At maturity, females remain brown and males turn black. Females stay near their burrows throughout their lives. Males, on the other hand, begin a wandering behavior from July to November, when they are frequently seen crossing highways. They are in search of nonwandering, mature females of any age. Few males survive the year they reach maturity since they die a natural death or are killed by the female during courtship or after mating. Females live 17 years or more.

This longevity of 12 years for males and 17 years or more for female tarantulas is rare because both young and adults are prey to several animals. Young spiders are consumed by birds, lizards, frogs, toads, snakes, and a species of parasitic fly. Adult tarantulas fall prey to a species of wasp, rodents, skunks, coatimundis, and raccoons. However, predatory mammals find tarantulas an uneasy catch because tarantulas defend themselves by using the "cloud of hair" trick. The tarantula accomplishes this feat by removing toxin-coated abdominal hairs from its abdomen with rapidly vibrating legs, creating a cloudy-hair mist in the air. If this mist contacts the mucous membranes of the eyes or the nose of a sniffing mammal, it experiences great discomfort. This discomfort gives the tarantula a slight advantage to escape.

Although behavior and size make the tarantula interesting, size may be detrimental because large spiders are more visible to "spider-fearing-folks." Some say, "The only good spider is a dead spider." It is difficult to convince these individuals otherwise, even with the analogy that follows. The little puppy is furry and so is the tarantula. The tarantula has eight legs; the puppy has four. The bite of both animals is painful. In the long run, the tarantula may be the safer of the two. The reason—tarantulas will not transmit rabies, carry blood-sucking ticks, or serve as a haven for disease-infested fleas.

BIBLIOGRAPHY

Comstock, J. H. 1975. The spider book. Ithaca, NY: The Comstock Publishing Company.

Gertsch, W. J. 1949. American spiders. New York: Van Nostrand.

Preston-Mafham, R., and K. Preston-Mafham. 1984. Spiders of the world. New York: Facts on File Publications.

MODIFICATIONS OF STEMS
AND LEAVES

Remember stories from Uncle Remus? Brer Rabbit begged Brer Fox and Brer Bear to do anything, except toss him into the brier patch. Brer Fox tossed Brer Rabbit into the brier patch anyway because it was there that Brer Rabbit was supposed to experience severe pain from thorns on brier bushes. Even today some dictionaries state that brier is any prickly or thorny bush of the rose family, including roses, raspberries, and blackberries. However, brier bushes are not thorny. Instead, roses and related berry shrubs of the brier group have prickles.

Prickles are sharp-pointed structures which grow from irregular places along a stem. Prickles are present on roses, blackberries, raspberries, gooseberries, and acacia. On the other hand, thorns are branches, sometimes with leaves, which grow only from buds situated in the angle between the petiole or stalk of the leaf blade and a stem. Thorns are stout, leafy, woody structures which terminate in a hard, sharp point. Plants with thorns include honeylocust, hawthorn, plum, osage orange, and pyracantha.

Spines differ from thorns in that spines are modified leaves instead of modified stems. Spines are the short and sharp structures on cacti. One particular cactus, the prickly pear, has truly been misnamed. The green, cylindrical, flat pad (a stem) of this cactus is incorrectly called a pear, and the words "prickly pear" imply "full of prickles." The barberry leaf has spines on its edges, and in some cases, the entire barberry leaf is a spine. Spines of black locust and mesquite are modified stipules rather than leaves. Stipules are paired structures situated at the base of the leaf's petiole or stalk. Spines on the desert plant, the ocotillo, are of special significance. Each is a persistent midrib. A midrib extends from the tip of the leaf blade along its center to the petiole.

Tendrils are also modified stems and leaves. Tendrils wrap around structures and assist the plant in vertical growth. Tendrils in Boston ivy, Virginia creeper, and grapevines are modified stems. Tendrils in the garden pea are extensions of the ends of leaves.

Cladophylls are modified stems which are leaf-like and extend outward from the edible portion of the asparagus plant. Cladophylls on asparagus are very fine in structure and resemble feathers fanning out from the tail of a peacock.

All of the above structures deviate in function from typical leaves and stems. For example, cladophylls increase surface area of the stem to maximize sun exposure, and tendrils aid upright growth to a plant which lacks a sturdy support stem. Prickles, thorns, and spines serve as defensive structures which reduce predation by herbivores (plant-eaters).

BIBLIOGRAPHY

Harris, J. D. 1880. Uncle Remus the story of Brer Fox and Brer Rabbit. London: Thomas Nelson and Sons, LTD.

Raven, P. H., R. F. Evert, and S. E. Eichhorn. 1992. Biology of plants. 5th ed. New York: Worth Publishers.

Torrey, R. E. 1932. General botany for colleges. New York: Appleton-Century-Croft, Inc.

TECHNIQUES OF
PREY SURVIVAL

Tony Gennaro
2 95

Interaction between a prey animal and its predators is serious. A victorious prey stays alive; a defeated prey dies. However, just as a predator has means to detect, attack, capture, and consume, a prey animal displays countermeasures to those procedures.

Here are examples in which prey avoid detection. Some species of prey may match background colorations of their habitat. Others, such as sea gulls, remove fragments of eggshells from nest sites to prevent predator detection of newly hatched offspring. As soon as they finish feeding on leaf blades, certain species of caterpillars snip through petioles or stalks of the partially eaten blades, causing them to fall to the ground. Such behavior reduces detection of caterpillars because certain kinds of predacious birds search for partially eaten leaves hoping to find a critter behind them.

If prey are detected, they act to prevent an attack. For example, when bees build a honeycomb on low vegetation where ants are foraging, bees secrete a substance around the supporting branch of the honeycomb between it and the ants. This substance is a barrier and prevents ants from snatching honey. In another example, gazelles "stott" once they are detected by a cheetah. Stotting means a gazelle continues to run a short distance and jumps about two feet off the ground with all four legs held stiff and straight. During each jump, the gazelle exposes its white rump patch to the predator. Stotting apparently informs the cheetah that an attack is no surprise, and therefore, capture is unlikely.

Now what action would prey take if a serious attack were underway, and the distance between predator and prey were diminishing rapidly? Belding's ground squirrels, which maintain high densities in mountain meadows, have a solution. The first squirrel to detect an attacking hawk or falcon utters a high-pitched whistle to warn other squirrels in the area. Immediately following its warning call, the squirrel heads for cover. Without delay, other squirrels respond to the warning call by seeking cover.

If all fails and a prey is captured, chances are great the prey will be consumed, unless, of course, it can persuade its captor to let go. Some species do just that. Black widow spiders jab sticky strands of silk into the face of an attacking mouse, and a species of larval moth sprays its attacker with a strong-scented aerosol of acid. Does the latter technique remind us of a last-minute evasive action we might take?

BIBLIOGRAPHY

Alcock, J. 1993. Animal behavior. 5th ed. Sunderland, MA: Sinauer Associates, Inc.

TERMITES

Yell, "Termites!" and people will run in all directions. And there is good reason for it. These tiny, colorless insects have wood-digesting microorganisms in their gut which are capable of turning mansions into a pile of chemicals which plants use to construct their own protoplasm. Termites have an important function in the ecosystem; they recycle nutrients from wood, and they have been doing that for 100 million years.

In North America, these tiny recycling critters live in underground (subterranean) nests with galleries connecting to a food source in contact with the ground. Such food sources may include dead trees, fence posts, or timbers of a house. Termites also dwell in nests entirely above-ground in dry or damp wood. Their homesite depends on the species of termite and its geographic distribution. For example, dry-wood termites inhabit the South, particularly South Carolina to Texas. Damp-wood termites are common in Florida, but they also inhabit western states. Subterranean termites are widely distributed.

Organization of different species of termites varies, but in general, a colony consists of several different castes. A caste is composed of individuals which perform specific tasks. One caste includes a reproductive pair which produces eggs. These eggs develop into other castes within the colony. One caste consists of workers. These termites are pale, soft, eyeless creatures which care for eggs and young. Workers construct, repair, and clean the nest, and they provide food for other castes. Another caste consists of soldiers which are blind and have large heads with long mandibles (jaws) for defense. These mandibulate soldiers are replaced in some species with soldiers which have underdeveloped mandibles (nasute soldiers) and a head drawn-out into what appears to be a long pointed nose (oilcan-shaped head). The liquid secreted from the tip of this pointed structure is used for defense. This secretion is also used to cement materials together, such as dirt, excrement, and partially-digested wood used in the construction of walls of the nest. Some developing termites mature into pigmented, winged males and females which are not a separate caste. These individuals have functional sight and emerge from termite nests annually. An emerging termite of one sex from one colony mates with the opposite sex emerging from another colony. The offspring of this pair become the colony of a newly established termite nest, the members of which will possess a variety of new genetic material.

These emerging winged forms are the only termites which see the light of day. Other members are confined to the darkness of tunnels. Obviously, communication must exist among castes in order for them to coordinate functions. One type of communication among termites consists of the use of a variety of chemicals, called pheromones. One termite, for example, emits a pheromone which initiates a response from another termite. Another method of communication involves the response of termites to light or air currents which enter a breach (tear) in the nest wall during an attack by ants or other predators. Vibrations of the nest during such an attack initiates jerking body movements among soldiers, which in turn rock their heads upwards and downwards several times a second. With each upward thrust, forelegs are lifted off the floor, and the termite heads are slammed against the ceiling of the nest. The resulting sounds are transmitted throughout the nest. These jerking motions are repeated by other soldiers who perceive the banging noises. These sounds, plus sensitivity to light and air cause some termites to retreat to deeper parts of the nest, especially those individuals in the nursery chambers. Other termites respond by attempting to repair and defend at the site of the attack. Soldiers attack with open mandibles or in other species, nasute soldiers squirt a sticky repulsive fluid. This fluid repels the invaders and at the same time immobilizes them. Alarmed termites make pheromonal trails which guide termites to areas of retreat, repair, and defense.

All these coordinated efforts of communication are the termites' means to save the colony. If minimal damage occurs during the attack and termites are successful in their repair and defense, termites which retreated to safe quarters, return to their original sites. If the invasion is massive, and the colony is destroyed, survival of the termite species depends on colonies developed elsewhere by the previous dispersal of winged reproductives.

BIBLIOGRAPHY

Borror, D. J., and R. E. White. 1974. A field guide to the insects. Boston: Houghton Mifflin Company.

Howse, P. E. 1970. Termites: a study in social behavior. London: Hutchinson and Company, LTD.

Swain, R. B. 1965. The insect guide. Garden City, NY: Doubleday and Company, Inc.

Weesner, F. M. 1965. The termites of the United States. Elizabeth, NJ: The National Pest Association.

Wilson, E. O. 1975. Sociobiology. Cambridge, MA: The Belknap Press.

MEMORY IN BIRDS

I remember visiting Aunt Hester in northern New Mexico during the early fall of last year. As she walked out to greet me, we chatted for a moment and then turned our attention to a bird feeder on her front porch where scrub jays were eagerly removing seeds from large fruiting heads of sunflowers. With seeds in mouth, the jays launched from the feeder and flew away. "Busy little characters," I commented. She responded, "Do you suppose those birds are storing the seeds somewhere? They keep returning time after time for more." My answer was, "Well, Aunt Hester, since the birds aren't marked, we really don't know if the same bird is returning, but yes, they could be storing those seeds. And, you know, those birds may remember the exact storage sites. In fact, I recall a bird study in which investigators demonstrated that birds can locate previous storage sites by memory."

This study involved a Clark's nutcracker. If this species relies on memory, the bird is quite impressive because it stashes as many as 9,000 caches of pine seeds (usually 1-10 seeds per cache) over hillsides during the fall. The bird digs a hole for each cache and then covers the cache with dirt. During the following winter and early spring, the nutcracker retrieves the seeds. Does the nutcracker use the sense of smell, vision, specific landmarks near the buried cache, or signs of digging and covering as cues to locate the caches of seeds? To answer these questions, a Clark's nutcracker was allowed to store seeds in a large outdoor aviary. Then, the bird was moved to another cage. The location of each cache was mapped, and all buried seeds were removed. Cache sites were cleared of nearby features and swept to remove signs of digging. Once released in the experimental aviary a week later, the nutcracker dug into as many as 80% of its previous cache sites while digging rarely in other places. These results suggest that the Clark's nutcracker does remember locations where it stores seeds.

Aunt Hester seemed pleased with my story. We never discussed the scrub jays again during that visit. I guess Aunt Hester and I both knew that although we were unaware of studies supporting memory in scrub jays, that the confident behavior of those birds demonstrated they would likely find their buried seeds during the following winter and early spring.

BIBLIOGRAPHY

Alcock, J. 1993. Animal behavior. 5th ed. Sunderland, MA: Sinauer Associates, Inc.

Balda, R. P. 1980. Recovery of cached seeds by a captive *Nucifraga caryocatactes*. Zeitschrift fur Tierpsychologie 52:331-346.

Tony Ferraro
95

DUCKBILL PLATYPUS

The duckbill platypus is a strange looking animal. In fact, in 1798, scientists at the British Museum examined the first platypus sent to them from Australia with disbelief. These scientists thought someone was playing a joke on them. That original specimen is still in the museum, complete with evidence showing that suspicious individuals attempted to locate ways in which someone attached the duckbill to the body.

The duckbill is a distinctive feature of the platypus's external appearance. Other features considered slightly strange because they accompany a bill include a furry pelt, webbed feet, and flat tail similar to that of a beaver. The female lacks nipples for feeding the young, although it does have mammary glands, and the male lacks external testes. Adult males also display a hollow spur about three-fourths inch in length on the ankles of each hind foot. This spur is used to inject venom into an opponent, usually a male platypus during conflict over space.

The internal anatomy of the platypus is also different from other mammals. Adult females lay eggs. Both sexes as adults lack teeth and have a cloaca which is a common sac that receives urinary, reproductive, and digested products before these materials are discharged from the body. The cloaca is present in reptiles and birds, but absent in all mammals except spiny anteaters which are close relatives of the platypus.

This strange looking mammal also rears its young in an interesting way. Ten days following mating, the female builds a brood nest of wet leaves and grass in a tunnel. This tunnel has its entrance above waterline along banks of a river or lake. The exact methods of laying eggs, incubating, hatching, and caring for the young still remain unknown to science; therefore, comments about these activities are based on assumptions. About two weeks after breeding, the female usually lays two eggs. On rare occasions, she lays one or three eggs. Incubation is thought to take at least ten days. Hatchlings are about one inch long, blind, and naked. These newly hatched platypuses make their way through the mother's fur to glandular areas on the belly where a milk-like substance, which is usually a fatty discharge, is secreted from mammary glands through pores in the skin. Apparently, the young platypuses lick this discharge from the fur or lap up the discharge from around the milk glands. After about 4 months, the fully furred, young platypuses, each about 14 inches in length, emerge from the burrow. These young platypuses attain a total length of about 17 inches.

Peculiar features and behaviors of the platypus mentioned above are enough to consider the platypus unique among mammals; however, in 1986 a group of scientists discovered another feature which sets the platypus and the spiny anteater apart from other mammals. This feature is the presence of electroreceptors in the bill of the platypus which detect electric currents emitted by contracting muscles of shrimp, fishes, earthworms, and insect larvae. These organisms are prey of the platypus, and their electrical impulses enable the platypus to detect and capture them. Such findings support suggestions from scientists that there are senses in animals yet to be discovered.

BIBLIOGRAPHY

Gregory, E. May 1991. Tuned-in, turned-on platypus. American Museum of Natural History 5:30-36.

Heckner, U. 1990. Egg-laying mammals (monotremes). Pages 194-203 in Grzimek's Encyclopedia of mammals. Vol. 1. New York: McGraw-Hill Book Company, Inc.

Hoffman, E. January-February 1990. Paradoxes of the platypus. International Wildlife 20:18-21.

Nowak, R. M. 1991. Walker's mammals of the world. 5th ed. Vol. 2. Baltimore: The Johns Hopkins University Press.

Voelker, W. 1986. Natural history of living mammals. Medford, NJ: Plexus Publishing, Inc.

Tony Gennaro 95

MY EXPERIENCE WITH AN
AMERICAN KESTREL

Come on Hank. Go for it. This is your chance for freedom." Those were my words to a little falcon, an American kestrel, which was perched on a tree next to our home in Portales, NM. Hank simply looked down at me, preened himself, and if he could speak, I am certain his words would have been, "I go where you go."

My story about Hank began in June 1962. At that time I was conducting fieldwork in the Chihuahuan Desert for my doctoral dissertation. Bill, my field assistant, had just graduated from high school. We were compatible, except for one thing. In the evenings after work, Bill would spend most of his time on a nearby hillside talking to his favorite companion, a prairie falcon. In those days, falcons, hawks, and owls were not protected by federal and state laws. Bill's time on the hillside bothered me. After all, it does get lonesome in the middle of nowhere, and I would have been happy to chat with someone during evening hours. Bill noticed my discontent.

While returning to the field following a weekend at our homes, Bill presented his solution to the problem. Bill directed my attention to a small cardboard box in his hands. "I've got a present for you," he said. As I was driving, Bill opened the box and pulled out an American kestrel hatchling. He held the bird in his outstretched hand and said, "This little fellow was abandoned by his parents. He's yours." I named the falcon Hank. The two of us began our companionship that day.

I liked that little falcon from the very beginning, and he seemed to return my affection, but maybe he had no choice. Hank was imprinting on me. Imprinting is a behavior which associates an animal with a resource. Birds commonly imprint. At this particular time in his young life, Hank would have imprinted on any nearby moving thing (living or nonliving). However, I felt that Hank's bond with me was more than imprinting, since Hank enjoyed my company and showed no fear of me. I enjoyed being trusted by such a small, helpless animal, and it made me feel good when Hank wagged his tail up and down each time I approached him.

To make sure Hank stayed with me, I applied a thin, flexible, three-inch leather jess to each of his legs and attached a long leash to the free ends of the jesses. This is a common procedure for restraining falcons in captivity. From that day forward, Hank's leash served as his lifeline to me.

Hank and I got along well in the field during the summer of 1962. He never left my side. When I was inside or outside my vehicle, Hank perched on my shoulder. We talked often. After all, prospectors talk to their donkeys, and cowboys talk to their horses. And, sometimes it seemed as though Hank answered me, at least with his eyes.

I was amazed at Hank's ability to learn. At mealtime, I fed him small pieces of lean stew meat which I wrapped in aluminum foil and kept in an ice chest. Hank quickly made two associations—one involved the presence of meat with the sound of the ice chest lid closing, and the other with meat and aluminum foil. When he heard the sound of the lid or observed foil of any kind, even foil on chewing gum, he would respond with a "klee, klee, klee," as he wagged his tail up and down.

One of Hank's instinctive behaviors, however, irritated me, or in plain words—pushed me beyond the limit. When Hank observed a raptor flying overhead (even an airplane), he would emit his warning call—"killy, killy, killy," and then he would flutter and fly all over the vehicle's interior and me while still on his leash. Obviously, he was attempting to escape from the raptor. My comments were, "Hank, stop it, stop it." Finally, he would settle down, and we would continue on our way. But, one time I really lost my cool and uttered words more harsh than, "stop it." That incident occurred on a very hot afternoon when I was attempting to cross a wide, sandy arroyo. I had the vehicle in low gear and full throttle. Sweat was flowing down my forehead, and my sweating hands were gripping the steering wheel with all my strength as the vehicle's rear end jumped up and down and from side to side. A reduction of power, and the vehicle would have been axle-deep in sand. At that time, Hank observed a raptor flying above us. Hank screamed, "killy, killy, killy," and began his typical escape reaction—tied to a leash. Those were trying moments.

During the next several years, Hank was also my companion out of the field. He was in my home or offices at the University of New Mexico, St. John's University in Minnesota, and Eastern New Mexico University. While in my offices, Hank stayed in a very large cage or flew freely about the room. All went well in those offices except during Hank's breeding seasons. At that time, Hank continuously wanted to mate with my hand or my head. In Hank's mind, I was his mate. Also, he would continuously emit, "klee, klee, klee" as he tried to persuade me to enter his cage. Hank typically made that noise with a piece of meat in his mouth as he stood on the floor of the cage or on his perch. Hank was using the meat to entice me. As he called, he would continuously bob his head up and down. At other times, Hank shredded paper in his cage to build "our" nest, meanwhile screaming, "klee, klee, klee." During those times, I found it very difficult to concentrate on work.

In the field or at home, one thing was certain. I was Hank's trusted companion. Hank never allowed anyone near his cage or in a room when I was not present. He would fly at the wire mesh on his cage when anyone else approached. This startled many individuals who approached his cage. If anyone entered the room, other than me, when Hank was out of his cage, he launched from his perch on my desk and dived straight at them. My three daughters (now grown and with their own families) tell how they teased Hank in his cage at home when I was not there. They would purposely go near Hank's cage so he would fly at them. Even today, one daughter comments, "I hated that bird." Now my three daughters admit they were jealous of Hank because of the attention I gave him.

However, as long as I was in Hank's presence, he never attacked anyone. In fact, for two or three years, Hank was star of an educational television show entitled, "Mr. Hank." During taping of those shows, Hank sat on his perch next to my guests and did not attack a single one of them.

During all my experiences with Hank, I never regretted having him. He was my faithful friend. But then, one day, or maybe subtly on several days, I thought perhaps possession of Hank was a selfish act. I began to feel very guilty because I had created Hank's world. He never had a chance to experience a relationship with his peers. I finally decided to give Hank his freedom. He was four years old at that time.

Hank's release did not go as planned. I removed his jesses and placed him on a fence near our house. Then, I walked indoors. I turned around and there he was with his talons in the screen of the door. The next evening, I walked over to a large tree near the house. I placed Hank on a low branch and sat near him. I talked to him periodically until darkness approached. He responded by wagging his tail. Then, knowing he would not fly in the dark, I carefully walked into the house. That was the longest walk of my life. The next morning, Hank was not in the tree. But I did observe him flapping his tail up and down as he perched on a telephone wire above me. He had captured a horned lizard, and he was letting me know by screaming, "klee, klee klee." All went well for one more day, then Hank, in his own way, included our residence in his protective realm. No one was allowed near our house except me. His response to an intruder, including other members of my family, was a direct attack. I had no choice but to walk up to Hank as he perched on a fence, place my hand out to him, and say, "Hop on, Hank." In my mind, Hank had made his choice--me. I had no other option but to keep him. He knew no other life except the one I had taught him.

One morning, fourteen years and two months after Hank was given to me, I walked in my usual way into the living room to greet him. My little friend lay stiff and stretched out on the floor of his cage. This time, Hank did not wag his tail and utter his cheerful welcome sound-- "klee, klee, klee." I buried Hank in the backyard of our house under a grave marker of cement upon which I wrote, "Hank, June 1962-August 1976."

I mourned the loss of my companion and the fact that I had deprived him of a life in the wild. I have been told several times that Hank had a good life. Perhaps. But my own opinion, then as well as now, is if humans create domesticated breeds, humans are responsible for the welfare of those breeds. Wildlife, on the other hand, should not be carelessly removed from their natural settings.